Energy and Transport Processes

Presentations at the 2008 AIChE Annual Meeting

Philadelphia, Pennsylvania
16 - 21 November 2008

ISBN: 978-1-61567-222-6

Printed from e-media with permission by:

Curran Associates, Inc.
57 Morehouse Lane
Red Hook, NY 12571
www.proceedings.com

Some format issues inherent in the e-media version may also appear in this print version.

Printed by Curran Associates, Inc. (2009)

AIChE
3 Park Avenue
New York, NY 10016-5991

Phone: (203) 702-7660
Fax: (203) 775-5177

www.aiche.org

TABLE OF CONTENTS

49a Dual Function Inhibitors for Methane Hydrate

Chongwei Xiao, Department of Chemical & Petroleum Engineering, University of Wyoming, Dept. 3295, 1000 E. University Ave., Laramie, WY 82071 and Hertanto Adidharma, Department of Chemical & Petroelum Engineering, University of Wyoming, Dept. 3295, 1000 E. University Ave., Laramie, WY 82071.

The formation of gas hydrates in oil and gas industries have been the subject of long-standing problems. For example, the hydrate formation may occur and block gas pipelines, which can lead to safety hazards. It may also occur in the drilling fluids that are used in deep offshore drilling operations, resulting in severe threats towards the operation safety. All of these also lead to catastrophic economic losses and ecological risks.

Various methods have been developed to control and mitigate the formation of gas hydrate. The techniques include heating, depressurization, water removal, and inhibition. In many cases, however, the hydrate inhibition by adding inhibitors is the only viable option. Two types of inhibitors are used nowadays: thermodynamic and kinetic inhibitors. Since exploration and production moves to deeper seas, the temperature and pressure in the field become in favor of hydrate formation and the addition of thermodynamic inhibitor becomes expensive and environmentally prohibitive; thermodynamic inhibitors need to be used at high concentration up to 50 wt%. The shortcomings of the thermodynamic inhibitors stimulated the search for kinetic inhibitors, which retard the hydrate formation by slowing down the hydrate nucleation rate, growth rate, and/or agglomeration rate. Kinetic inhibitors can be effective at low dosage (<1%) and therefore have significant economic and environmental advantages. The existing kinetic inhibitors, however, are still not believed to give an economic solution especially at high pressure and large degree of supercooling. It has also been identified for some cases that the combination of thermodynamic and kinetic inhibitors is still needed to give better results. Therefore, there is still a need to discover inhibitors that are more effective than the existing inhibitors.

The goals of this work are therefore to discover a new class of novel inhibitors. The performance of imidazolium-based ionic liquids as a new class of gas hydrate inhibitors has been investigated. Their effects on the equilibrium hydrate dissociation curve in a pressure range of 30 to 110 bar and the induction time of hydrate formation at 106 bar and a high degree of supercooling, i.e., about 25oC, are measured in a high-pressure micro differential scanning calorimeter. It is found that these ionic liquids, due to their strong electrostatic charges and hydrogen bond with water, could shift the equilibrium hydrate dissociation/stability curve to a lower temperature and, at the same time, retard the hydrate formation by slowing down the hydrate nucleation rate, thus are able to act as both thermodynamic and kinetic inhibitors. This dual function is expected to make this type of inhibitors perform more effectively than the existing inhibitors.

49b Kinetics of Asphaltene Precipitation: A Novel Perspective

Tabish Maqbool, Iman Aden Hussien, and H. Scott Fogler. The University of Michigan, 2300 Hayward St., Ann Arbor, MI 48109-2136

This research investigates the kinetics of asphaltene precipitation from crude oils using n-alkane precipitants. Conventionally, it has been understood that the precipitation of asphaltenes is governed by the amount of precipitant added to the crude oil. Recent literature shows that there is a kinetic phenomenon associated with asphaltene precipitation. In this research we have demonstrated that depending upon the amount of n-alkane precipitant added, the time required to precipitate the asphaltenes may vary from a few minutes to several months. Therefore, the onset of asphaltene precipitation is a function of the concentration of precipitant and time. A technique to quantify the amount of asphaltenes precipitated as a function of time and precipitant concentration is also discussed. The kinetic effects caused by various precipitants have been also investigated in this work. Refractive index measurements were that were made, provided further insight into the kinetics of asphaltene precipitation. In order to compare the nature of asphaltenes precipitated early in the precipitation process to the asphaltenes precipitated at later times, polarity based fractionation and dielectric constant measurements were used. Results from these experiments show that the asphaltenes precipitating at different times from the same crude oil-precipitant mixture are different from each other.

PARTIAL RECOVERY OF WAX GEL STRENGTH: QUANTITATIVE AND QUALITATIVE ANALYSES

Kyeongseok Oh, Mark Jemmett, Pankaj Tiwari
Jules Magda and Milind Deo
University of Utah, Salt Lake City, UT 84112

Abstract

Waxy components in a crude oil start to precipitate when the surrounding temperature is lower than wax appearance temperature (WAT). While the wax deposition can be initiated during the flow, wax gel formation occurs primarily under static conditions. When the wax gel develops within a relatively short time, certain pressure is needed to overcome the yield strength of the gel along the pipeline for restart. It was found that paraffinic components contribute to the evolving gel strength continuously while the oil is cooled below pour point (PP). Gel strength depends on wax amount and wax composition in the gel network. It has been reported that the gel properties depend on various factors: temperature, cooling rate, cooling time, shear history, and a diverse combination of factors. This study explores gel strength by stress exertion below the PP followed by further cooling. Model oil used in this study was prepared by mixing mineral oil and well-characterized wax. The measurements of WAT and PP were performed using the ASTM methods. A controlled-stress rheometer equipped with a cone-and-plate geometry and a Peltier plate device was employed to determine the yield stress and the measurement of creep response. The cooling was scheduled after applying stresses in the creep range. Yield behavior was compared after applying varying stresses and cooling to lower temperatures.

Key words: wax, gel strength, yield stress, creep compliance

Introduction

High molecular paraffinic waxes in a crude oil start to precipitate when the surrounding temperature is lower than the wax appearance temperature (WAT). Pour point is another important characteristic temperature usually determined by ASTM D97 (1), which represents a transition from flow to no flow. Flow discontinuity occurs in due to either wax deposition or wax gel formation. While the wax deposition can be initiated during flow, wax gel formation occurs primarily under static conditions caused by either planned or emergency shutdown. When the wax gel develops within a relatively short time, certain level of pressure application upstream is necessary to overcome the yield stress of the gel along the pipeline for restart (2, 3). Various rheological studies of gelled waxy oil have been published. Boger and coworkers (4, 5) discussed the existence of three definite characteristic responses, which they categorized as elastic, creep and fracture when the gel was subjected to shear. In the elastic region, the gel strength is fully recovered after sufficiently low shear is applied to the gel regardless of duration time over which the shear is applied. Prior to the static yield stress (the point at which the gel fractures), a creep region is observed, in which the gel strength is partially recovered once the applied stress is released. The gel yields at shorter times, when higher shear is applied. In case of fracture, the gel breakage occurs when the loading stress is high enough to deform the gel network.

Singh et al.(6) pointed out that various factors, such as wax/oil ratio (wax amounts), molecular weight of the wax, cooling rate and mechanical shear history, affect wax precipitation and deposition characteristics including gelation temperature. In particular, they reported that when shear is applied to waxy oils, the gelation temperature is depressed, with a greater reduction at higher shear rates. They also compared the effect of cooling rates on the gelation temperatures, showing that slower cooling rates resulted in greater reductions in yield stresses. Venkatesan et al. (7) examined the yield stresses with different cooling rates in both quiescent and shear conditions. Highest yield value was observed in quiescent condition with slowest cooling rate. Under shear application, the yield stress increase was observed with the highest cooling rate. Lopes da Silva and Coutinho (8) explained that the gelation measurement is obtained from the initial network development with low level of cross linking.

Building weaker gel would be favorable to the restart process. Deo et al. (9) presented that the restart pressure was decreased when they applied the certain pressure during gel formation in a laboratory-scale flow loop. Ekweribe et al. (10) showed that restart pressures decreased when increasing pressures were applied during gel formation. Both results were obtained by pressure induction into the pipeline at temperatures higher than WAT. The pressure application at a temperature above WAT suppresses the gel formation if the pressure induces mechanical shear on the gel. The yield strength of the gel being cooled with shear is important for restart considerations.

In this study, we examine the gel strength behavior by measuring a static yield stress at different temperatures and creep test combined with secondary cooling.

Experiment

Model oil

Model oil was prepared by using a well-characterized wax and white mineral oil. The carbon number distribution of wax is shown in Figure 1. This study used 5 wt.% wax in oil.

WAT and Pour point

Modified ASTM methods were used to determine WAT and pour point using a temperature-controlled bath and a cell-type jacket for coolant circulation. Temperature in the cooling bath was set to 10 °C, 0 °C and -5 °C to determine both the WAT and the pour point. The measurement was carried out at 3 °C intervals at first and then in narrower temperature intervals as the measurements got closer to the target. The pour point was defined as the temperature 1°C higher than the temperature of no-flow observation. The pour point measurement was repeated. The results were presented in Table 1.

Rheometer measurement

A controlled-stress rheometer (AR 500, TA instruments, Inc) equipped with a cone-and-plate geometry and a Peltier plate device was employed to determine the yield stress and the measurement of creep response. Cooling was scheduled after various applications of stresses in the creep range. Yield stress was determined by the ramping stress rate at 1 Pa/sec. Gap setting was 48 micrometer and cone angle is 2°.

Results and Discussion

Yield stress and creep response

Yield stress was determined by stress ramping rate of 1 Pa/sec. The sample was heated to 50°C in an oven and was loaded in the rheometer with the plate at 35°C. In the rheometer, the sample was heated to 55°C heating for 10 minutes, and the cooling run was begun. Cooling rate was fixed for all measurements. The static yield stress values of 195-215 Pa at 5°C and 115-135 Pa at 10°C are observed. Creep yield was also examined with different creep stresses at 5°C and 10°C. Compliance responses measured at 5°C are shown in Figure 2 and creep yield values are presented in Table 2. Lower stress values resulted in longer time to break the waxy gel as expected.

Yield stress after creep application at same temperature

Yield stress was determined after creep stresses were applied until the creep compliance response reached a certain value that was chosen before the gel fracture occurred. Creep compliance showed three different regions as shown in Figure 3. Gel breakage or fracture was observed by cone rotation after the compliance values increased exponentially. Additional creep recovery test confirmed that the gel strength recovered partially after creep stress was released before gel fracture occurred. This study examined the creep effect on the static yield stress. Table 3 presents the yield stress results after creep stresses, 150 Pa and 100 Pa, applied differently. In instances when stresses were applied to the gel until the compliance value reached 1e-4 (1/Pa) with 150 Pa and 1.4e-4 (1/Pa) with 100 Pa, the static yields were 205 Pa and 199 Pa. These values were within the measured yield stress values of 195-215 Pa at 5°C. However, with the creep compliances of 2e-4 (1/Pa) in 150 Pa creep stress, 1.6e-4 (1/Pa) and 2.4e-4 (1/Pa) in 100 Pa creep stress, lower static yield values of 190 Pa, 187 Pa and 168 Pa were obtained. Lower yield stresses were obtained after creep stress was applied to higher compliance values at same temperature, which indicated a gel strength loss.

Combined effect of creep application and secondary cooling

The gel strength was examined by secondary cooling combined with different magnitude of creep stress compliance. Figure 3 shows the creep compliance during creeping the gel at 90 Pa at 10°C. Creep stress of 90 Pa was applied to the gel (10°C) differently in the compliance range of 1e-4 to 5e-4 (1/Pa). Creep applied gel at 10°C was cooled down again to 5°C before yield stress measurement. Yield values obtained are presented in Table 4. In the cases of creep application in the compliance range of 1e-4 (1/Pa) to 4e-4 (1/Pa) at 10°C, the static yield values measured at 5°C were higher than the yield stress values without creep application (247-278 Pa with creep versus 195-215 Pa without creep). The creep stress contributes to the increase of compactness of the initial gel network resulting in stronger gel formation after secondary cooling. Other case of creep compliance up to 5e-4 (1/Pa), the beginning of exponential increase, resulted in lower yield value (172 Pa) than the yield without creep. This compliance value, 5e-4 (1/Pa) (see Figure 3) resulted in the partial loss of gel strength, which resulted in less static yield value at 5°C. Similar observations were made when creep stress was applied at 15°C. The gel strength increase at 5°C (266 Pa) was observed in case of creep stress with 50 Pa up to 9e-4 (1/Pa) compliance. The compliance values recorded before gel fracture increased at lower temperature. In other cases of the creep stress application just before the cone rotation of rheometer, 50 Pa up to 1e-2 (1/Pa) and 30 Pa up to 1e-3 (1/Pa) at 15°C, the yield stress values at 5°C were much lower (83 Pa and 51 Pa). As a reference value, the yield stress at 5°C was 10-15 Pa after gel fracture by cone rotation at 15°C.

Conclusion

Gel strength was examined by measuring the static yield stress, creep response and yield stress combined with secondary cooling after creep stress application. The static yield stress increased with decreasing temperatures. In creep test, yielding time was dependent on the magnitude of creep stresses. When the yield stress was measured after applying the creep stress at the same temperature, the yield stress values decreased after increasing the magnitude of creep compliance. Gel strength after creep stress application and secondary cooling was also explored. While the gel strength after secondary cooling was lower when the gel fracture initiated, the gel strength dramatically increased when stresses in the creep range were applied prior to secondary cooling. This study provides insight to various scenarios which may exist during gel formation along the pipeline. Previous studies confirm that the initial pressure induction during the gel formation resulted in lower restart pressures. However this study shows evidence of gel strength increases when the pressure is applied at temperatures below PP.

Reference

1. Annual Book of ASTM-Standards, "Petroleum Products, Lubrications. West Conshohocken, Pennsylvania: American Society for Testing and Materials," Sect. 5., 1999.
2. Davenport, T.C. and Somper, R.S.H. (1971), "The yield value and breakdown of crude oil gels," *J. Inst. Petrol.*, 57 (554), pp86-105.
3. Rønningsen, H.P. (1992), "Rheological behaviour of gelled, waxy North Sea crude oils," *J. Pet. Sci. Eng.*, 7, pp177-213.
4. Chang, C., Boger, D.V. and Nguyen, Q.D.(1998), "The yielding of waxy crude oils," *Ind. Eng. Chem. Res.,* 37, pp1551-1559.
5. Wardhaugh, L.T. and Boger, D.V.(1991) "The measurement and description of the yielding behavior of waxy crude oil," *J. Rheol.*, 35 (6), pp1121-1156.
6. Singh, P., Fogler, H.S. and Nagarajan, N.(1999), "Prediction of the wax content of the incipient wax-oil gel in a pipeline: An application of the controlled-stress rheometer. *J. Rheol.*, 43, pp1437-1459.
7. Venkatesan, R., Nagarajan, N.R., Paso, K., Yi, Y.-B., Sastry, A.M. and Fogler, H.S.(2005) "The strength of paraffin gels formed under static and flow conditions," *Chem. Eng. Sci.*, 60, pp3587-3598.
8. Lopes da Silva, J.A. and Coutinho, J.A.P.(2004), "Dynamic Rheological Analysis of the Gelation Behavior of Waxy Crude Oils," *Rheol. Acta*, 43, 433-441.
9. Deo, M., Guimaraes, K.O.R., Oh, K.; Magda, J.J., Venkatesan, R. and Montesi, A. (2007), "Flow Assurance of Wax Gelled Oil," The 8th International Conference on Petroleum Phase Behavior and Fouling, Pau, France, June 10-14.
10. Ekweribe, C., Civan, F., Lee, H.S. and Singh, P. (2008) "Effect of System Pressure on Restart Conditions of Subsea Pipelines" SPE 115672 presented at the 2008 SPE Annual Technical Conference and Exhibition, Denver, CO, Sep. 21-24.

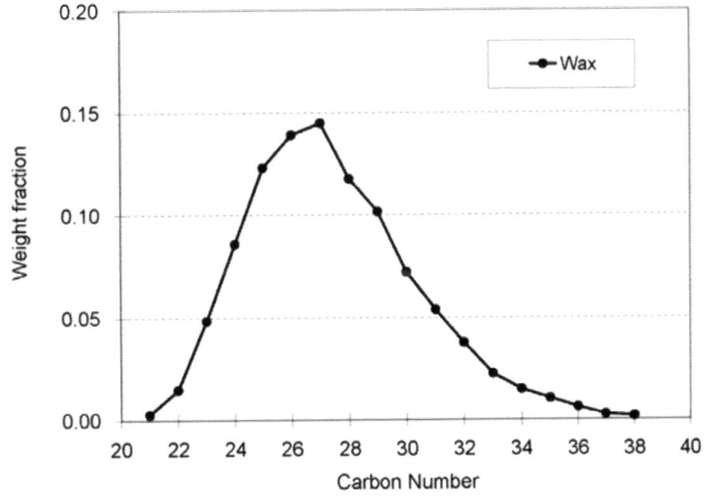

Figure 1. Carbon number distribution measured by SIMDIS.

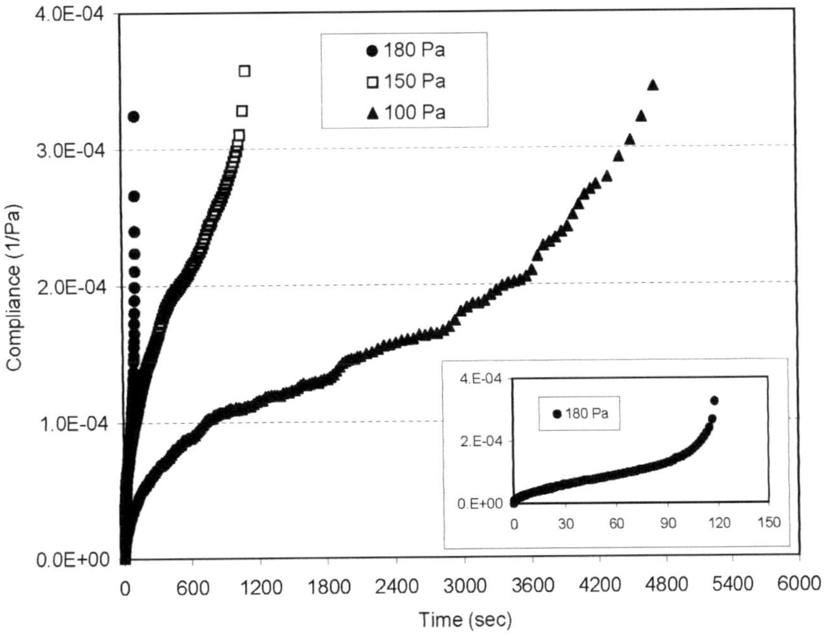

Figure 2. Compliance behaviors when the various creep stresses were applied at 5 °C.

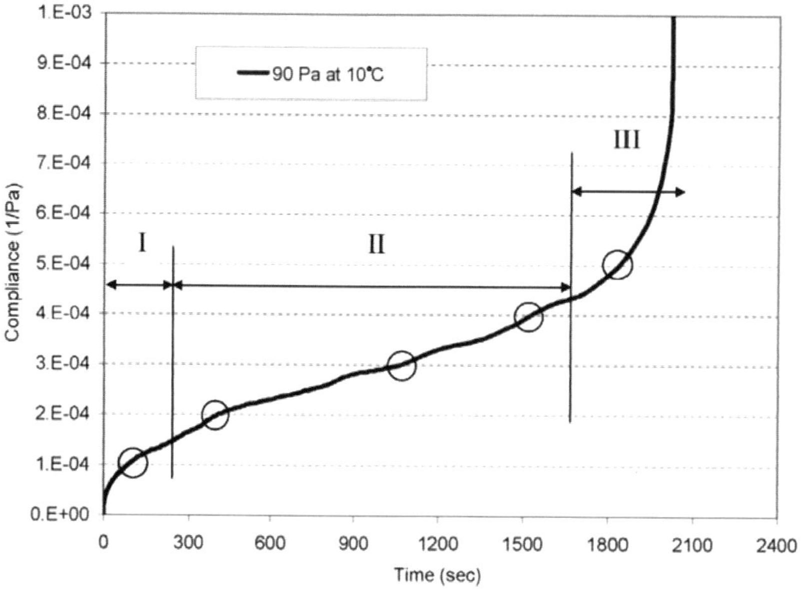

Figure 3. Compliance response when the creep stress of 90 Pa was applied to 5 wt.% wax gel at 10 °C.

Table 1. WAT and PP data of 5 wt.% wax model oil measured by modified ASTM methods (D2500 and D97) at different temperatures.

Cooling bath temperature	10 °C	0 °C	-5 °C
WAT	31 °C	34 °C	35 °C
PP	21 °C	22 °C	23 °C

Table 2. Creep responses of 5 wt.% of wax model oil. Longer time was required when the less creep stress was applied at certain temperatures. Deviation values were added when the tests were carried more than two times.

Temperature	Creep stress	Time to break the gel
5 °C	180 Pa	2 min
	150 Pa	20 min
	100 Pa	100 ± 10 min
10 °C	100 Pa	4 ± 1 min
	90 Pa	34 min
	70 Pa	59 min

Table 3. Yield stresses of 5wt.% wax gel measured at the ramping rate of 1 Pa/sec after creep stresses were applied until the certain compliance values were reached at 5 °C

Creep stress	Creep compliance (1/Pa)	Ramping yield after creep
150 Pa	1e-4	205 Pa
150 Pa	2e-4	190 Pa
100 Pa	1.4e-4	199 Pa
100 Pa	1.6e-4	187 Pa
100 Pa	2.4e-4	168 Pa

Table 4. Combined effect of creep application and secondary cooling in various compliance values followed by ramping yield stresses

Initial Temperature	Creep stress	Creep compliance (1/Pa)	Decreased Temperature	Ramping yield
10 °C	90 Pa	1e-4	5 °C	260 Pa
10 °C	90 Pa	2e-4	5 °C	278 Pa
10 °C	90 Pa	3e-4	5 °C	247 Pa
10 °C	90 Pa	4e-4	5 °C	259 Pa
10 °C	90 Pa	5e-4	5 °C	172 Pa
15 °C	50 Pa	9e-4	5 °C	266 Pa
15 °C	50 Pa	1e-2	5 °C	83 Pa
15 °C	30 Pa	1e-3	5 °C	51 Pa

49d Flow Assurance: In-Process Particle and Droplet Characterization for Process Design and Optimization

Eric J. Hukkanen, Real-Time Analytics, Mettler-Toledo AutoChem, Inc., 14833 NE 87th Street, Redmond, WA 98052 and Benjamin Smith, Particle Systems Characterization - Process Analytical Technology, Mettler Toledo, Redmond, WA 98052.

There are several flow assurance problems occuring during oil production, recovery, and transport. In cases of extreme temperatures and pressures, it is possible that methane gas hydrates crystallize or asphaltenes precipitate in the pipeline. If not properly controlled, the hydrate crystals or asphaltene particles may agglomerate to the point of plugging the pipeline. This can be costly it deposition or blockage may damage the reservoir and affect the oil production rate. Therefore, it is critical to study how additives may be used to better understand, optimize, and control agglomeration and precipitation.

In the case of oil recovery and transport, water is used to displace oil in the well. Ultimately, the water and particulates must be removed from the oil, whether by physical or chemical methods. Therefore it is necessary to quantify how much water or particulate remains trapped in the oil before and after the unit operation. For example, de-emulsifying studies are conducted to improve the separation process of oil and water, by varying the type and concentration of different additives.

Probe-based Lasentec tools are ideal for characterizing methane gas hydrates, asphaltene particles, and oil emulsions. In most cases, the conditions (high pressure) to which hydrate crystals form and asphaltenes precipitate make it extremely difficult to sample. The particle and droplet systems are extremely sensitive to temperature and pressure fluctuations. Additionally, off-line techniques require that the samples are dilute enough to allow for simple measurement. However, the measurement of dilute samples is not representative of the actual system and provides a very limited data set. With Lasentec FBRM and PVM probes the researcher can track the entire dynamic particle and droplet system in crude oil at standard operating temperatures and pressures, without sampling or sample dilution.

This paper will present several flow assurance case studies in which Lasentec and PVM are implemented to characterize the formation of crude oil emulsions, hydrate crystals, and asphaltene particles.

106a Characterization of Oilfield Waxes Using Nuclear Magnetic Resonance Spectroscopy

Shuqiang Gao, Ying Jiang, and Anthony Rodriguez. Champion Technologies, 3130 FM 521, Fresno, TX 77545

Keeping crude oil in pipelines flowing without interruption is of economic and strategic importance. Oils are complex mixtures and contain varying amounts of naturally occurring waxes. Because waxes can crystallize and stick to the wall inside the pipelines when the oil temperature falls below the wax appearance temperature, wax deposition is a significant challenge in flow assurance. It was noticed that certain waxes are stickier than others and are harder to treat. Because this behavior is not well understood, the effectiveness of the flow assurance programs is hindered and the billion dollar production assets are at risk. In this work, Nuclear Magnetic Resonance (NMR) Spectroscopy was employed to characterize the structures of the different types of waxes from various crude oils. Results demonstrated that the sticker waxes have branched carbon chains while the non-sticky waxes are mainly straight chain paraffins. This finding provides a molecular level basis for devising effective wax treatment programs.

106b Wax Deposition Modeling In Oil/water Two-Phase Flows

Zhenyu Huang, Chemical Engineering, University of Michigan, 3074 H.H. Dow Building, 2300 Hayward St, Ann Arbor, MI 48109 and H. Scott Fogler, The University of Michigan, 2300 Hayward St., Ann Arbor, MI 48109-2136.

Wax deposition is a crucial economic issue in subsea oil production. Deposition modeling using fundamentals of transport phenomena and numerical analysis not only establishes scientific insight into the deposition mechanisms, but also predicts deposit profiles for effective remediation such as pigging. However, most of the research on deposition modeling is still limited to the idealized single phase flow conditions. In this study, a CFD model is developed to study the wax depositions under oil/water two-phase stratified flow and dispersed flow. In stratified flow systems, the effect of water volume fraction on deposition is discussed. In dispersed water-in-oil flow systems, it is shown that the radial distribution of water plays an important role in wax deposition by affecting the mixture properties and the heat/mass transfer in the boundary layer.

106c The Role of Polydispersity on Gel Strength In Alkane Solutions

Michael Senra, The University of Michigan, 2300 Hayward St., Ann Arbor, MI 48109-2136 and H. Scott Fogler, Chemical Engineering, The University of Michigan, 2300 Hayward St., Ann Arbor, MI 48109-2136.

Wax deposition is an issue of extreme importance and interest to the oil industry. One of the major issues regarding the deposit itself is the strength of the resultant gel that form once crystallization occurs. Greater insight into understanding what characteristics can cause strong deposits can be helpful in the proper development of wax inhibitors and/or a pigging schedule. Using binary and ternary systems, this research has shown that polydispersity and the ability of two or more components to cocrystallize both have major effects on the gelation characteristics. Additionally, the gelation characteristics of a system can be greatly influenced by the abundance of the crystallizable material present, both the least soluble component that dictates the cloud point and those more soluble components.

106d Modeling of Heavy Oils Under Wide Range of Conditions with the Friction Theory

Sergio E. Quiñones-Cisneros, Instituto de Investigaciones en Materiales, Universidad Nacional Autónoma de México, Apdo. Postal 70-360 Cto. Exterior, Ciudad Universitaria., Mexico D.F., 04510, Mexico, Patsy V. Ramírez-González, Facultad de Química, Departamento de Ingeniería Química, Universidad Nacional Autónoma de México, Ciudad Universitaria, Mexico D.F., 04510, Mexico, Torben Laursen, VLXE ApS, Scion-DTU, Diplomvej, Building 376, Lyngby, 2800, Jefferson Creek, Chevron Petroleum Technology Corporation, Houston, TX 77082, and Ulrich K. Deiters, Institute of Physical Chemistry, University of Cologne, Luxemburger Str. 116, Cologne, 50939, Germany.

From Natural Gas to Heavy and Waxy Oils, the friction theory (FT) has been extensively applied to the accurate description of the rheological behavior of a wide range of reservoir fluids. The method works in a self-consistent manner; by taking advantage of fundamental thermodynamic models widely used in the oil industry (such as cubic or even SAFT-type equations of state) delivers an integrated approach extended to rheological properties.

The main application of the FT is within the area of flow assurance, which is essential for all aspects of the petroleum industry. The flexibility and accuracy that the FT models deliver can, therefore, provide essential evaluation tools for flow assurance as well as many other potential applications where viscosity is a key parameter. The FT models can be applicable to the full range of temperature and pressure conditions typically found in the oil industry (i.e. from low- to high-temperature and pressures up to around 100 MPa).

In this work, several representative examples covering wide ranges of conditions are studied. Starting from the basic standard information that is found in regular laboratory reports, full thermodynamic and viscosity modeling is achieved. Subsequently, the stability and consistency of the developed models is shown even beyond tuning conditions. In essence, the approach has two possible modes of application: 1) prediction and 2) tuning. The use of generalized parameters can deliver results that go from the accurate viscosity predictions of light fluids to reasonable predictions in the case of heavy oils having viscosities in ranges as high as thousands of cP. Alternatively, if viscosity data is available highly accurate FT models can be derived. The FT models can also be combined giving sensible predictions for oil blends or even for applications involving strong compositional changes such as VAPEX processes.

The present work demonstrates how thermodynamic fluid models can be used for the accurate modeling of not only typical thermodynamic properties but also transport properties and viscosity in particular. The approach links fundamental thermodynamics to viscosity in a unique way. This work will, therefore, present an integrated alternative for the modeling of a number of thermophysical properties that are highly relevant to engineering developments in the oil and gas industry applicable to a wide range of conditions as well as fluids.

Kinetics of Oil Shale Pyrolysis and Oil Composition

Milind D. Deo and Pankaj Tiwari, Chemical Engineering, University of Utah, 50 S. Central Campus Dr., MEB, Rm. 3290, Salt Lake City, UT 84112.

Abstract

The oil shale industry is going through a revolution of sorts. After the oil crisis in the 1970s, a great deal of effort was spent on research and development and on pilot technologies. The oil price collapse in the early and mid eighties led to the total discontinuation of the oil shale research and development programs. Extensive research was conducted both in on-surface and in-situ production methods. Even though some large pilot underground retorting operations were performed, the on-surface (mining and processing) methods were closest to full-scale (~10,000 barrels/day) commercial implementation. The in-situ production technologies have seen a significant revival since some recent advances. The slow thermal pyrolysis of the organic matter in shale leads to a light oil product that would not require additional thermal upgrading. In this paper we present isothermal and non-isothermal kinetics of the pyrolysis of the Greenriver shale using thermo gravimetric analyses.

Introduction

Oil shale can be a viable alternative source for fast depleting natural sources of petroleum. The organic part in the oil shale undergoes chemical decomposition on thermal heating, known as pyrolysis and a liquid substance (shale oil) is produced. The yield and quality of the pyrolysis products depend on source material (geological availability, quality, size of the rock and its constituents), temperature, pressure, and time of the reaction. The main constituent of this organic part of the shale is Kerogen (which can be represented approximately as $C_{200}H_{300}SN_5O_{11}$, ~10 wt/wt % of rock shale), and small percent of bitumen is present. The extent of kerogen decomposition, yield of the products and kinetics of the reaction also depend on the environment of thermal heating, like in the presence nitrogen (pyrolysis) or air (combustion).

In this study, the kinetics of the Mahogany oil shale are obtained using thermal gravimetric analysis (TGA, Q-500). The present study covers isothermal (300-600°C) and non-isothermal (0.5 to 50°C/min) decompositions of crushed Mahogany Oil Shale (single particle size, 100 mesh) with TGA in N_2 (Pyrolysis) and air (Combustion) environments. Data obtained from TGA are analyzed in detail. Different mathematical approaches for both the cases are applied to determine the kinetic parameters (activation energy, Ea and pre-exponential factor, A) with the assumption of first order reaction. Comparisons between pyrolysis and combustion kinetic data, and with the data in the literature, are included in this work. For isothermal experiments, the data were fit using the integral method and the activation energy and pre-exponential factor, for the N_2 environment were 135 kJ/mol (100 kJ/mol for air) and 1.2E+09 (5.1E+07 for air) respectively. Non-isothermal kinetic data were derived using four different methods, namely, direct Arrhenius plot, integral method, Friedman approach and the maximum rate method. The goodness of the fit and values of kinetic parameters are different, and depend on the mathematical method used for the same sample. Data and detailed analyses are provided in this extended abstract. The tradeoff between activation energy and pre-exponential factor is also considered in obtaining a good fit.

Experimental Section- TGA

The mahogany oil shale samples were crushed to 100 mesh size and then dried for four hours to remove moisture, if any was present (there was no weight loss found during drying in the TGA experiments). The TGA experiments were performed with 20-30 mg of oil shale and other operating conditions such as purging time (10 min) and flow rate of purge gas (60ml/min) were kept fixed for all experiments. For the isothermal experiments, the thermal induction time was kept as low as possible (100°C/min, the maximum ramp rate of TGA Q-500 and 5 hour reaction time). The non-isothermal experiments were performed to 1000°C (the highest temperature possible in TGA Q-500). All of the experiments were repeated and found to be reproducible.

Kinetic Expression- Mathematical models

Different mathematical approaches were used to determine the kinetic parameters (activation energy, Ea and pre-exponential factor, A) with the assumption of first order reaction. The kinetics of the pyrolysis of Mahogany oil shale were studied for different, constant heating rates and constant temperature profile. Tables 1-4 show the experimental conditions used for both the cases. Some of the non-isothermal and isothermal experiments were done with N_2 and Air environments.

The kinetic parameters were determined using the following simple mechanism.

Kerogen $\longrightarrow$ Products

The decomposition profile of kerogen in the air environment gives two peaks for both isothermal and non-isothermal cases. Thus two successive first-order (series) reactions were used to model and analyze the data. The reaction is assumed in the form,

Kerogen $\longrightarrow$ Bitumen $\longrightarrow$ Products

Where, bitumen is the intermediate product.

As some pre-test experiments with TGA (up to 1000°C) indicate, 10-12 wt % of the total shale is organic (assumed to be kerogen here). Thus all the data were normalized on the basis of 10 wt/wt % of the kerogen. The conversion of kerogen to product is defined as,

$$\alpha = \frac{(W_\circ - Wt)}{(W_\circ - W\infty)} \quad \text{or} \quad \alpha = \frac{(W_\circ - Wt)}{(W_\circ * X)} \qquad (1)$$

Where, $W_0 =$ Initial Weight of the sample, (mg), W_t = Weight of the sample at time t, (mg), $W_\infty =$ Weight of the sample at the end of the experiment, (mg), X = Percent of the organic part, decomposing during experiment time (max 10%).

With first-order reaction, the rate law is expressed as follows.

$$\frac{d\alpha}{dt} = A.\exp(\frac{-Ea}{R.T}).(1-\alpha) \qquad (2)$$

Where, A = Frequency (Pre-exponential) factor, (S^{-1}), Ea = Activation energy, ($KJ.mol^{-1}$), R =Gas constant, (8.314 KJ. $mol^{-1.}K^{-1}$), T =Temperature, (K).

The value of X is not fixed in this work but, used as the actual value. In the isothermal experiments, the thermal induction period is not included in the analysis of data, thus the decomposition of the organic continues during this period. The conversion data were normalized from the isothermal condition to the end point of the analysis. Non-isothermal TGA data were analyzed in the range of organic weight loss. Conversion profiles for each non-isothermal experiment were normalized from zero to one. The temperature at which the weight derivative begins to rise is chosen as the zero conversion point and the temperature at which weight derivative returns to the base line is the end point (complete conversion of the organic part) of the analysis.

Different methods were employed to study the kinetics of the Mahogany oil shale. For the isothermal data, the integral method was employed, and for non-isothermal analysis, four methods (differential, integral, Friedman, and maximum rate method) were used.

Isothermal: Integral method

The integral form of the equation (2) can be written as

$$\ln(1-\alpha) = -k.(t - t\circ) \qquad (3)$$

Where, t_o is the time at the start of the constant- temperature period (when isothermal condition is reached). In this study, the thermal induction part is eliminated from the kinetic analysis and correspondingly the W_∞ is corrected.

Non-isothermal: Differential (Direct Arrhenius plot) method

The kinetic rate expression for non-isothermal experiments with Arrhenius dependency can be derived by introducing heating rate (β) in equation (2)

$$\frac{d\alpha}{dT} = \frac{A}{\beta}.\exp(\frac{-Ea}{R.T}).(1-\alpha) \qquad (4)$$

Where, $\beta = \frac{dT}{dt}$

$$\ln\left[\frac{1}{(1-\alpha)}\right]\frac{d\alpha}{dT} = \ln(\frac{A}{\beta}) - (\frac{-Ea}{R.T}) \qquad (4)$$

The plot of $\ln\left[\frac{1}{(1-\alpha)}\right]\frac{d\alpha}{dT}$ against 1/T will give the starlight line, and values of *Ea* and A can be obtained by this line.

Non-isothermal: Integral method

Overall conversion vs. temperature curves

$$\int_0^\alpha \frac{d\alpha}{(1-\alpha)} = \frac{A}{\beta}\int_{T\circ}^{T} \exp(\frac{-Ea}{R.T})dT$$

Finally equation can be rearranged as,

$$\ln\left[\frac{-\beta.\ln(1-\alpha)}{RT^2}\right] - \ln(1 - \frac{2RT}{Ea}) = \ln(\frac{A}{Ea}) - (\frac{-Ea}{RT}) \qquad (6)$$

The value of Ea and A can be obtained by a least square fit of the equation (e) to the experimental data. This can be done either iteratively or directly. The simplified form of the integral method used is,

$$\ln\left[\frac{-\ln(1-\alpha)}{T^2}\right] = \ln(\frac{A.R}{Ea.\beta}) - (\frac{Ea}{RT})$$

Non-isothermal: Friedman method

Friedman procedure assumes successive first order reaction

$$\ln(\frac{d\alpha}{dt}) = \ln\left[A(1-\alpha)\right] - (\frac{Ea}{R.T}) \qquad (7)$$

Non-isothermal data can be analysed for specific conversion point (α) for all different heating rate, and the plot for $\ln(\frac{d\alpha}{dt})$ verses 1/T will gives a straight line. The slope and intercept will give the activation energy and pre exponential factor respectively.

Non-isothermal: Maximum rate method

Maximum rate method is a mathematical tool, which explains the maximum rate at a point when following condition is satisfied,

$$\frac{d\alpha^2}{dT^2} = 0$$

Thus, for first order reaction (n=1), equation (4) gives:

$$\ln(\frac{\beta}{T_m^{\ 2}}) = \ln(\frac{A.R}{E} - (\frac{Ea}{R.T_m}) \qquad (8)$$

Where T_m, is the temperature at the maximum reaction rate and α_m will be the conversion at that condition.

T_m can be calculated from weight loss curves with different heating rates for the same sample. Apparent activation energy Ea and apparent frequency factor A can be determined by linear slope and intercept, respectively, equation (8).

Results and Discussion

TGA Curve

The total extractable kerogen content in Mahogany Oil Shale is about 10 % of the total weight. There was no significant weight loss during pre heating observed which confirmed the absence of moisture in the sample. This was confirmed in TGA experiments, where no peak detection and weight loss was seen below 150°C, Figure 1. The reproducibly of the data was observed in all experiments, for example, Figure 1. There are two derivative peaks in all non-isothermal experiments, corresponding to organic and carbonate decompositions. The carbonate decomposition commenced at 525°C or above depending upon the heating rate and gave a weight loss of about 25-30%. There was a shift in maximum rate toward higher temperature with the increase in heating rate from 0.5 to 50°C/min. This difference is due to shorter exposure time to particular temperature at faster heating rates Figures 2 and 3.

In the nitrogen environment, only one peak is present in the range of the 300-550°C indicating that only one reaction or process occurs in this temperature range. While in air environment there are two peaks in the same temperature range, which indicate there are two reactions occurring simultaneously.

Two possible mechanisms were tested

1. A single step mechanism involving, kerogen → Products (N_2)
2. A two step mechanism represented as kerogen → Bitumen → Products (Air)

Isothermal

For isothermal experiments, the data were fit using the integral method. The activation energy and pre-exponential factors, for N_2 environment were 135 kJ/mole and 1.2E+09 respectively, while in air environment these were 100 kJ/mole and 5.1E+07 respectively.

Non Isothermal

Non-isothermal kinetic data were derived with four different methods, namely, direct Arrhenius plot, integral method, Friedman approach and Maximum rate method. The goodness of the fit, and the values of kinetic parameters are different, and depend on the mathematical method used for the same sample.

Conclusion

The kinetic mechanism of organic decomposition is different in different environments (N_2 and air). Fast heating provides a shift in decomposition rate curve towards higher temperature. Activation energy is more in N_2 than in air environment for the isothermal case. For non-isothermal experiments, different method gives different values for kinetic parameters. The best fit tradeoff relationship between activation energy and pre exponential factor for N_2 is found with Friedman procedure while for air with maximum rate method.

References

1. "Kinetics of the thermal decomposition of oil shale from Puertollano (Spain)", M.C. Torrente, M.A. Galan, Fuel, 80, 2001, 327-334

2. " Kinetics of Pyrolysis of Moroccan oil Shale by Thermogravimetry", Deepak S. Thakur and H Eric Nuttall, Ind. Eng. Chem. Res. 26, 1987, 1351-135

3. Study of different kinetic models for oil shale Pyrolysis, Shuyuan Li, Changtao Yue, Fuel Processing technology 85, 2003, 51-61

433

5. "Study of Pyrolysis of Oil Shale", Shuyuan Li, Changtao Yue, Fuel, 82, 2003, 337-342.

Table: 1

ISOTHERMAL					N_2_normalized data	INTEGRAL APPRAOCH			
T	Total time	Initial weight	Start_ Isothermal temp		Normalization factor *X*	N2		Slope	
°C	min	mg	Time	Weight Loss		1/T	R^2	K	lnK
300	720	26.758	3.33	99.17	0.0917	0.00174	0.5163	0.0008	-7.1309
350	240	26.69	3.82	98.75	0.0875	0.0016	0.8893	0.0045	-5.40368
400	240	22.643	4.36	98.21	0.0821	0.00148	0.9644	0.0311	-3.47055
450	240	24.688	4.96	96.01	0.0601	0.00138	0.8157	0.2973	-1.21301
500	240	25.001	5.59	**89.29**					
550	180	23.953	6.3	**88.38**		These data can not be analyzed as before reaching Isothermal condition they lost more than 10% weight ,which corresponding to organic part in shale			
600	30	24.101	6.96	**87.89**					

Table: 2

ISOTHERMAL			Air_ normalized data				INTEGRAL APPRAOCH		
T	Total time	Initial weight	Start_ Isothermal temp		Normalization factor *X*	Air		Slope	
oC	min	mg	Time	Weight Loss		1/T, K	R^2	K	lnK
300	240	23.643	2.7	99.27	0.0927	0.00174	0.873	0.0365	-3.31044
350	240	23.391	3.82	98.14	0.0814	0.0016	0.8752	0.1909	-1.65601
400	240	23.241	4.31	96.85	0.0685	0.00148	0.8455	0.8725	-0.13639
450	180	32.162	4.88	94.32	0.0432	0.00138	0.864	2.8006	1.02983

Table: 3

NONISOTHERMAL						N2_Single Step mechanism						DIFFERENTIAL METHOD	
Heating rate	Initial weight	Start		End		max			R^2	slope	Intercept	Activation energy	A
Beta	mg	T	wt %	T	wt%	Tmax	wt%	%/min					
0.5	22.645	255.66	98.68	421.6	91.98	392.76	93.52	0.039	0.961	9351	10.8	77.744214	24510
1	28.642	269.6	98.84	437.6	92.52	398.33	94.21	0.062	0.959	9002	10.07	74.842628	23624
2	26.94	280.06	98.67	456.4	91.57	414.15	93.48	0.181	0.962	10379	11.8	86.291006	266505
5	25.972	348.93	97.83	474	90.59	432.29	92.83	0.074	0.977	14873	17.93	123.65412	3E+08
10	38.452	349.79	98.26	490	90.33	445.62	92.74	1.348	0.972	14905	17.54	123.92017	4E+08
20	29.493	371.61	98.42	504	89.32	460.18	92.08	3.549	0.979	17757	21	147.6317	3E+10
50	22.374	377.32	98.57	530.6	88.87	477.03	92.11	9.099	0.967	17218	19.56	143.15045	2E+10

Table: 4

NONISOTHERMAL						Air _ TGA data								
		First Peak							Second Peak					
Ramp rate	Initial weight	Start		End		max			End		max			weight loss
Beta	mg	T	wt %	T	wt%	Tmax	wt%	%/min	T	wt%	Tmax	wt%	%/min	%
0.5	18.688	179	99.33	311.3	91.84	279.9	94.8	0.11	396.4	87.09	340.2	89.2	0.108	12.91
1	20.261	199.1	99.41	323.8	92.05	294.1	94.71	0.1103	400.39	86.98	354.42	89.26	0.107	13.02
2	19.981	201.9	99.51	339.1	92.44	305.9	95.2	0.097	421.5	87.71	367.1	90.07	0.094	12.29
5	30.562	211.4	99.56	358.5	92.54	323.8	95.32	0.4545	459.65	87.28	392.22	89.87	0.437	12.72
10	34.986	216.5	99.98	374.9	92.31	337.1	95.18	0.9074	499.49	86.87	409.59	89.82	0.747	13.13
20	21.695	215.5	99.62	389.2	92.42	341.1	95.9	1.697	504.6	86.93	425.93	89.87	1.445	13.07
50	30.223	227.7	99.58	395.3	93.19	351.4	95.97	3.466	522.99	86.97	450.45	89.8	3.902	13.03

Figure: 1 TGA Curve -Reproducibility of Data (Two Peaks)

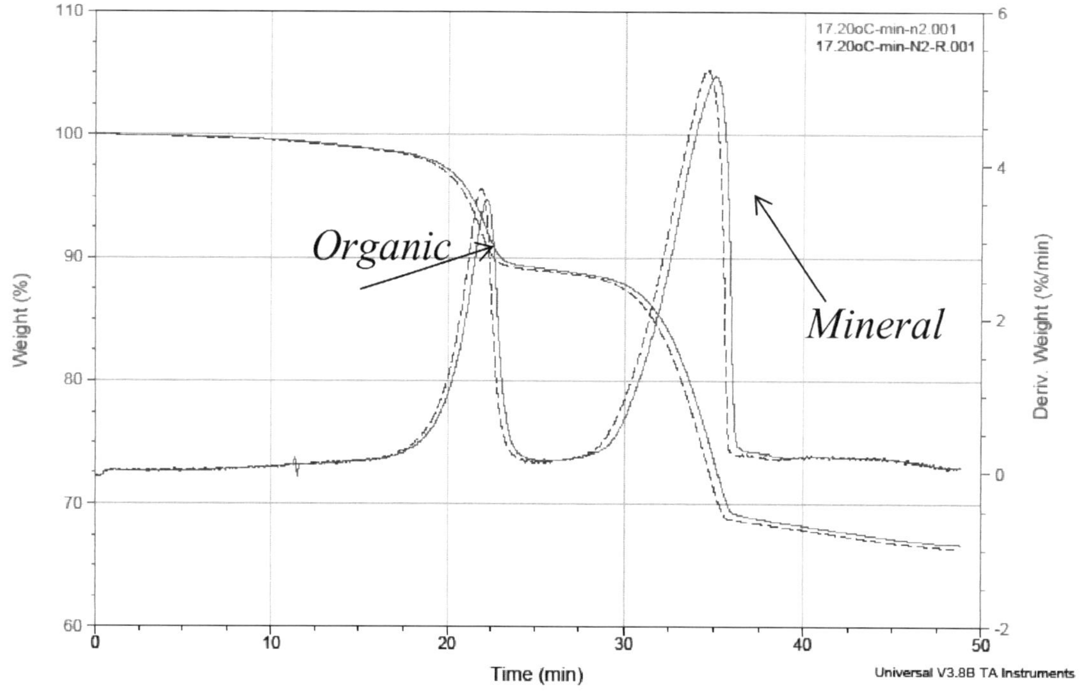

Figure: 2 Non- isothermal data in N_2 environment (Single step mechanism for organic part)

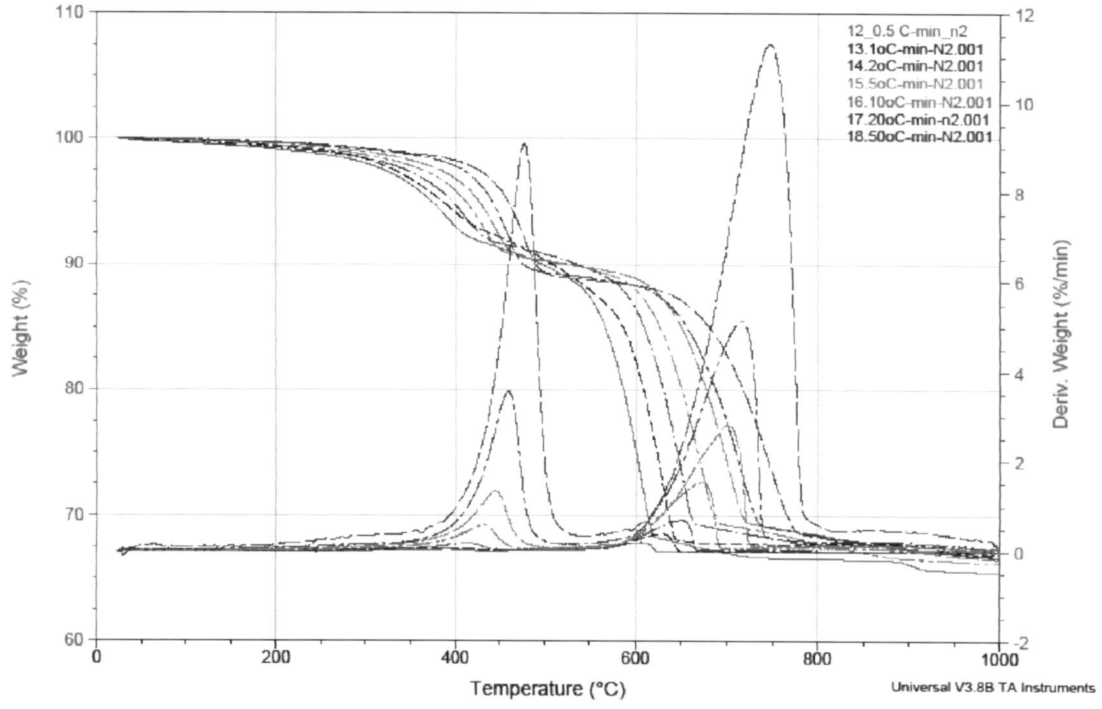

Figure: 3 Non- isothermal data in Air environment (two step mechanism for organic part)

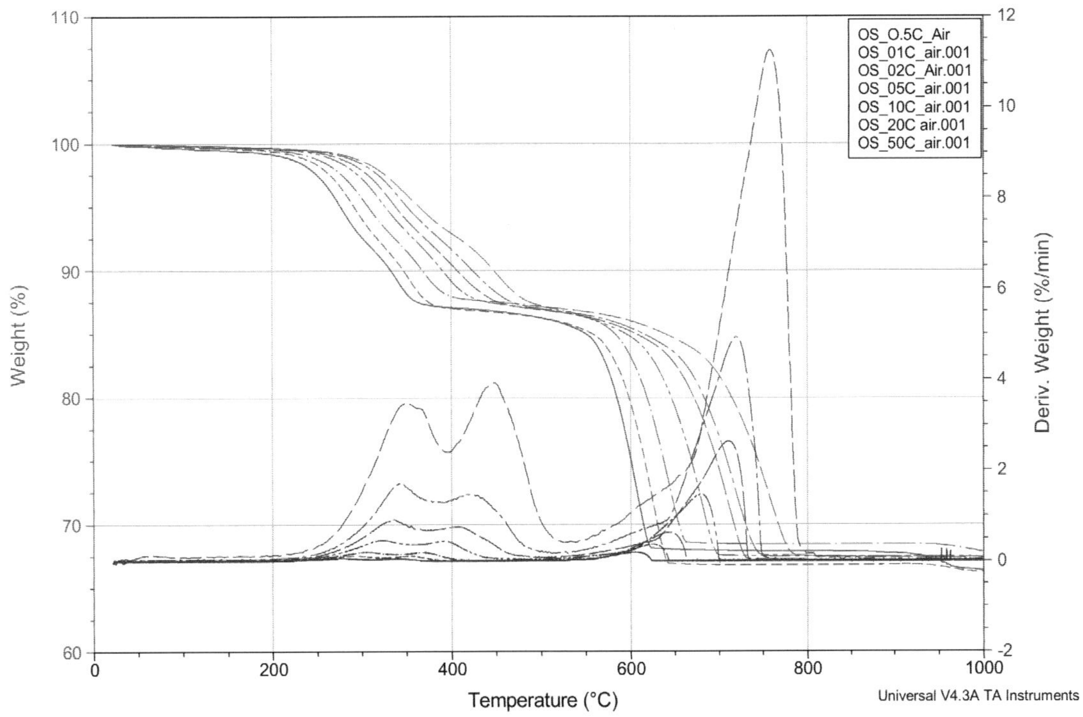

172b Heavy Oil Combustion In Porous Media and Fractured System

Gerald Debenest, Institut de Mécanique des Fluides de Toulouse, allée du Professeur Camille Soula, Toulouse, France, Hossein Fadaei, IMFT, Allée du professeur Camille Soula, Toulouse, 31400, France, Alexandre Lapene, G.E.M.P, Institut de Mécanique des Fluides de Toulouse (I.M.F.T) / TOTAL, 1 allée du professeur Camille Soula, Toulouse, 31400, France, Michel Quintard, Institut de Mécanique des Fluides, allee C. Soula, Toulouse, 31400, France, and Arjan Kamp, Uppa, Chloe, Bâtiment IFR Rue Jules Ferry, Pau, France.

The aim of this paper is to present the preliminary results obtained in two studies performed at IMFT Lab. regarding in-situ combustion of heavy oil. We are working on EOR (Enhanced oil recovery) in different geological systems containing a significant amount of heavy oil.

One concerns the water effects on oil combustion. The experimental work performed has shown an important effect of water in the vapor phase on some reactions and this is not taken into account in general in the reactive scheme used.

The second study is about combustion in fractured system. A

characterization of the front propagation has been made by varying the transport coefficients (such as effective diffusion coefficient in the matrix part of the medium and permeability in the fracture zone). The results show that combustion process is mainly governed by diffusion in the matrix.

Some multi-blocks simulations will be given.

172c Modeling In-Situ Production of Oil from Oil Shale

Milind D. Deo, Chung-Kan Huang, and Pankaj Tiwari. Chemical Engineering, University of Utah, 50 S. Central Campus Dr., MEB, Rm. 3290, Salt Lake City, UT 84112

Production of oil from oil shale reservoirs requires thermal input. In the laboratory, temperatures of about 350 C are seen as necessary to achieve sufficient conversion of kerogen to oil. In in-situ operations, the heat can be imparted to the medium in the form of a convective medium (steam or any other suitable gas) or by resistive heating. The temperature profiles in the reservoir are governed by the rate and mode of the heat transfer method. Once the reservoir begins to be heated, the oil generation and production process is governed by a set of highly complex coupled processes – heat transfer to the shale, kerogen decomposition, rock transformation and product transfer. In this paper, we explore the modes of heat transfer (convective and conductive) and the interplay of the time-temperature history in the reservoir with geomechanical and multiphase aspects of oil production from oil shale. The conductive heating mode reduces the need to create surface area for fluid injection – but, the energy requirements are significantly higher. Providing the heat through injected fluids is challenging in this impermeable medium – but provides an opportunity to explore innovative chemistry options and to lessen energy usage. There are technology options to make the oil production process energy neutral, and these are discussed in the paper.

172d The Production of Pickering Emulsions for Heavy-Oil Pipeline Transportation

Louis Fradette, Charles-Olivier Fournier, and Philippe Tanguy. URPEI, Chemical Engineering Department, Ecole Polytechnique de Montreal, P.O. Box 6079, Station Centre-Ville, Montreal, QC H3C3A7, Canada

For heavy-oil pipelining, an interesting alternative to dilution lays in the production of an emulsion of water-like viscosity. Pickering emulsions, or solid-stabilized emulsions, are generated with standard mixing equipment but do not require chemical surfactants while exhibiting extremely long shelf-lives and can be considered a valuable technological option. The fine solids positioned at the oil-water interface acts as the stabilizing agent and are thermodynamically more stable at this location than anywhere else if their affinity for the each phase is carefully selected. The rupture of the emulsion can be triggered by chemical or mechanical means instead of involving heat and consequently makes this type of emulsion very attractive from the economical perspective. In this work, the role of the dispersed phase viscosity in solid stabilized oil-in-water emulsions was studied using silicon oils of widely varying viscosities. Emulsification in controlled conditions showed that the viscosity of the dispersed phase acted as a damping factor for particle anchoring at the o/w interface. To our knowledge, this role had never been demonstrated. Furthermore, the emulsification results indicated the important combined effect of time and viscosity on the generation of a stable emulsion. It is demonstrated that, in the context of designing a continuous emulsification process for extra-heavy oil, the contact time is the dominating parameter.

172e The Filtration of Suspensions In Trickle Bed and Monolithic Reactors

M. Hamidipour[1], **F. Larachi**[1], and Zbigniew Ring[2]. (1) Chemical engineering department, LAVAL University, Quebec, QC G1K 7P4, Canada, (2) British petroleum, Residue Hydrocracking 150 W. Warrenville Rd., Naperville, IL T9G1A8

Canadian oil sand bitumen, equivalent to one-third of world's useful petroleum resources, is a promising alternative to counter declines in conventional crude oil sources. Upgrading, including hydrotreating, is necessary to comply with low sulphur specifications for fuels. Bitumen-derived feeds contain non-filterable ultrafines, typically <20 µm. These fines impose economic and processing setbacks in the downstream hydrotreaters where H2 and Heavy Gas Oil (HGO) flow co-currently downwards in a trickle bed in which the hydrotreating reactions remove objectionable materials. In the course of operation, plugging develops and leads to progressive increase in bed obstruction and pressure drop. The cumulative effect of tens of thousands of barrels of feed processed each day diverts the catalyst bed from its primary function to a huge filter. Fines accumulation causes the pressure drop to rise by restricting the flow. Eventually, the pressure drop becomes so high that the hydrotreaters are shutdown and the still chemically active catalyst is replaced. Fines deposition shortens hydrotreater cycle life, increases operational problems and maintenance work leading to poor overall energy efficiency (Gray et al., 2002). Therefore, research is needed to understand different features of this problem. In our group, the effect of gas flow was first studied on the extent of deposition by raising the gas superficial velocity from low to moderate values (Edouard et al., 2006). A decreasing trend of specific deposit (the mass of deposition per unit volume of the reactor) was observed. A correlation was established to relate the filtration rate and liquid holdup in the bed. Later on, the results of some experimental hydrodynamic observations were reported (Hamidipour et al., 2007). The complications to interpret the results of liquid circulation mode in comparison to the one-pass mode were discussed. An area was highlighted on the superficial velocity map of gas versus liquid to specify the conditions in which the trickle to pulse flow regime changeover is plausible under filtration conditions. The occurrence of plateau in pressure drop signal at the severe plugging conditions was attributed to the short circuits of fluids near the wall which consist of deposition islands and plug-free corridors.

In this communication, a cold-flow setup using kerosene-kaolin suspensions was studied with the goal to understand the deposition and release of fines in trickle bed reactors and in monolith beds. Iso-flow feed policies and cyclic operation (liquid, gas and alternating) were evaluated in terms of specific deposit, pressure drop and flow regime evolutions in the quest of the best approach for extending reactor operational life under filtration conditions. Electrical Capacitance Tomography (ECT), with the help of suitable reconstruction algorithms, was used as an aiding tool to track as a function of time the evolution of the plugging and release patterns inside the bed. The effects of bed length as well as gas and liquid superficial velocities was examined on the structure of deposition along the bed and recommendation are proposed regarding the most suitable feed policy to reduce deposition.

References:

-Gray M.R., Srinivasan N., Masliyah J.H. 2002. Pressure buildup in gas–liquid flow through packed beds due to deposition of fine particles. Can. J. Chem. Eng. 80, 346-354.

- D. Edouard, I. Iliuta, F. Larachi, 2006. Role of gas phase in the deposition dynamics of fine particles in trickle-bed reactors, Chem. Eng. Sci., 61, 3875.

-Hamidipour, M., Larachi, F., Ring, Z. 2007. Hydrodynamic observation of trickle beds under filtration conditions. Ind. Eng. Chem. Res. 46, 8336-8342.

218a Applications of a Micropyrolysis Index (MPI) towards Molecular Engineering of Fuels

Steven P. Crossley, Maria Pilar Ruiz, Liang Zhang, and Daniel E. Resasco. Cbme, University of Oklahoma, 100 East Boyd, Room T-335, Norman, OK 73019

A new sooting tendency measurement was developed to give an estimate of the pyrolytic sooting tendency of a particular fuel on a lab scale. This measurement has been coined the micropyrolysis index (MPI), as it involves pyrolyzing 20μ L of fuel across a hot bed of α-Al_2O_3 in a flow reactor in a He environment. The α-Al_2O_3 bed is then removed, and the deposited carbon is quantified via temperature programmed oxidation (TPO) in order to obtain a kinetic measurement of the fuel to form soot via pyrolysis. SEM, Raman, and same spot TEM images were used to characterize the deposited soot. The MPI compliments similar techniques, such as threshold sooting indices (TSI) or yield sooting indices (YSI), which introduce oxygen into the system. By comparing the two methods, one can separate the tendency of a compound to form soot based on pyrolysis alone with other factors that will vary from compound to compound such as flame temperatures and oxygen/fuel ratios.

Because such small sample volumes are used, several pure compounds were tested. A database was constructed, and quantitative structure-property relationships (QSPR) were created in order to predict the MPI of a pure compound based on its molecular structure. These measured and predicted values were used to provide direction for several important issues in catalytic conversion, including selective ring opening for diesel fuels, reduction of aromatics in gasoline while maintaining octane number, and enhancing fungibility of several renewable fuels.

DEEP DESULFURIZATION OF TRANSPORTATION FUELS WITH NANOREACTIONS

Li Zhou, Wei Dai, Shengqiang Wang, Yaping Zhou
High Pressure Adsorption laboratory
Tianin University, Tianjin, China

Introduction

Burning sulphur-laden transportation fuels results in misty and fuggy air of large cities. The combustion products of sulphur-laden fuels poison the exhaust conversion catalyst and invalidate the catalyst box. As results, the vehicle emission gets worse and worse no matter how advanced the auto technology is. Some sulphur compounds converted to sulphates after combustion and emitted into the atmosphere in the form of superfine particles. Suspension of these superfine particles and the inefficiently converted hydrocarbons yield misty and foggy atmosphere. Stringent regulations on the sulphur content of transportation fuels have been issued in developed countries, but not yet in developing countries. Nonetheless, deep desulfurization of transportation fuels has drawn the global research attention. Thiophene and its derivatives are the sulphur species most difficult to remove, and research on deep desulfurization is targeting at the removal of this kind of sulphur compounds.

The present state of deep desulfurization technology was extensively reviewed in literature.[1-4] Hydrodesulfurization is the technology practically applied. Most sulphur compounds such as thiols, sulfides, disulfides and a considerable portion of thiophenes were removed in hydrodesulfurization. However, some benzothiophenic compounds such as dibenzothiophenes and 4,6-dimethyldibenzothiophene (4,6-DMDBT) are still remained after the treatment. Hydrodesulfurization technology is not dependable for the target of deep desulfurization. Due to saturation of double bonds in olefin and/or aromatic components, excessive hydrodesulfurization of gasoline leads to decrease of octane number. The effectiveness of hydrodesulfurization decreases either for diesel following the increase of aromatic compounds and nitrides. In addition, deep hydrodesulfurization of diesels subject to constraints on fuel density, cetane number, and operation cost. Meanwhile, the catalytic activity decreases as a result of catalyst coking in deep hydrodesulfurization.[5,6]

The other desulfurization technologies reported in literature need further developments before commercialisation. These methods include oxidation, extraction, adsorption, alkylation and biodesulfurization. Adsorption and oxidation (oxidation-adsorption or oxidation-extraction) are considered promising among them for industry. The oxidation method changes sulphur compounds to sulfones or sulfoxides , and then the polarity difference with other components is utilized in a subsequent treatment. Different techniques were proposed for the oxidation method.[7-9] Catalysts are usually used in oxidation, and organic acids,[10,11] heteropoly acids and their salts,[12,13] and Ti-silica molecular sieve[14] were reported. The oxidant used includes 30% aqueous hydrogen peroxide[10,11,13,14], nitrogen dioxide,[15] oxygen[16] and air.[17] Polar organic solvents or ionic liquids were used as extractant.[18] The oxidation desulfurization method bears merits in mild operation condition and without expense of hydrogen. However, this method becomes disabled in fuels containing too much aromatic components and/or dissolved water. Adsorption method can be applied with or without hydrogenation, and the latter is classified into physical or chemical adsorption. The S-Zorb SRT (Sulfur Removal Technology)[19] is a representative of the hydrogenation-based adsorptive desulfurization techniques. Although the method bears merits in less loss of octane number and relatively high desulfurization efficiency, it is not a cost efficient technology since the fuel has to be vaporized and reacted at 380-420 ^{o}C. The desulfurization method basing on physical adsorption seems not applicable due to the low selectivity for thiophenic compounds. Prominent success was reported in chemical adsorption, and the IRVAD (Irvine Robert Varraveto Adsorption Desulfurization) method,[20] the π-complexion method,[21,22] and the SARS (selective adsorption for removing sulphur) method[23] are well known representatives. The IRVAD method is suitable for many liquid hydrocarbons and more than 90% sulphur can be removed. However, the sulphur capacity of adsorbents is low, and the adsorbents need to be frequently regenerated. The sulphur capacity of adsorbent was considerably increased relying on the principle of π-complexion. However, the adsorption performance declines greatly where aromatic compounds, oxidants and moisture are present in fuels.[24] The SARS method[4,23] preloads transition metals on an adsorbent, for example, Ni/SiO2-Al2O3, and reached the same or little higher sulphur capacity in comparison with the π-complexion method. The alkylation method is to increase the boiling point of sulphur compounds through alkylation reaction and separate sulphur via a subsequent operation. This kind of methods is limited by competitive reaction of aromatic components and the polymerisation of olefins.[25] Biodesulfurization might have industrial potential, but a lot of fundamental developments are required for a better industrial performance.[2,26,27]

The proposed method uses adsorbents, but the desulfurization function is not due to adsorption of sulphur compounds, but to conversion of them via a characteristic reaction occurring in the pore spaces of adsorbents. For the removal of thiophenic and benzothiophenic compounds, the condensation reaction with formaldehyde is characteristic. This reaction needs strong acid as catalyst, and different acids are suitable. The adsorbent may also have different choices in accordance with the catalyst. Since real fuels contain more sulphur compounds than thiophenic/benzothioiphenic, the total sulphur content of real fuels cannot reach the goal of deep desulfurization relying only on the condensation reaction. Therefore, oxidation was used as a succeeding characteristic reaction after removing thiophens.

Why does the characteristic reaction reside in adsorbents? First of all, it is to increase the probability of sulphur compounds contacting a pore/reactor; second, it is because the concentration of sulphur compounds is too low to have the reaction occurred in bulk phase, though the content of hundreds or thousands ppm is too high to be tolerant. When excessive quantity of catalyst/reagent was preloaded in the pore space of adsorbent, sulphur compound will change to another species of larger molecular weight and different property as soon as it touches the pore. Then the reaction product is adsorbed inside the pore. The large specific surface area/pore volume of the adsorbent greatly increases the probability of the touch. Finally, it is because the increase of catalytic activity in a reactor of nanometers dimension.[28]

Experimental

Adsorbent is the carrier of the reaction characteristic for sulphur compounds; therefore, it must load reagent and catalyst before desulfurization. The adsorbent must match with catalyst. For example, loading H_2SO_4 on carbonaceous material[29] and HCL[30] or phosphomolebdic acid (PMA) [31] on silica materials. Reagent of the condensation reaction is formaldehyde (FA), which is pre-loaded in adsorbent via vapor adsorption. H_2SO_4 or HCL may be then dropped in the adsorbent, while PMA may be loaded by soaking method before loading FA. Silica gel is used to load peracetic acid for the oxidation reaction, which is used for commercial fuels following the condensation reaction. The weight ratio of peracetic acid to silica gel was 0.4/1.0.

The desulfurization performance was firstly studied with model fuels and then tested in commercial fuels bought from a Sinopec gas station in Tianjin, China. Considering the difference in reactivity of different type of hydrocarbons, three types of model fuels were prepared. Octane was used to represent the aliphatic fuel (ALF) and benzene to represent the aromatic fuel (ALF). A mixture of 80 wt % octane and 20 wt % benzene to represent mixed type fuels (MXF). Since thiophenes and benzothiophenes constituted the obstacle of deep desulfurization, thiophene of 2000 ppm (equivalent to 760 ppm of sulfur element) and 4,6-DMDBT of 1000 ppm (equivalent to 151 ppm of sulfur element) were used to represent sulfur compounds. The sulfur content of thiophene in model fuels was detected by GC4000 gas chromatograph. The sample size was 2 μl and the minimum detectable sulfur content was about 1 ppm. The total sulfur content in commercial fuels varies time to time and is detected by a coulometric detector Model WK-2D.

The desulfurization condition was firstly searched in batch tests carried out in a flask with temperature control and agitation facilities. Then the testing fuel flows continuously passing through a packing bed at appropriate condition aiming to collect breakthrough curves, from which the sulfur capacity was determined. Dosage in a batch test is 20-30 g model fuel and 1-2 g adsorbent.[29-31] The sulfur content of fuel was analyzed intermittently. Breakthrough experiments were carried out in a vertical glass column (length 250 mm, diameter 10 mm) filled with the adsorbent at a constant temperature. The fuel was pumped up by a mini creep pump model BT01-YZ1515 and flowed down through the column at a constant rate. Breakthrough curves were drawn on plotting the transient sulfur content of fuel versus the cumulative fuel volume. The transient sulfur content was normalized with the initial sulfur content in feed while the cumulative fuel volume was reduced to unit mass of adsorbent. The reduced capacity of the adsorbent was calculated with following equations:

$$Q_b = (\frac{v\rho x_i}{m}) \times t_b \times 100\% \quad (1)$$

$$Q_s = (\frac{v\rho x_i}{m}) \int_0^{t_s} [1 - \frac{c_t}{c_i}] dt \times 100\% \quad (2)$$

Where Q_b is the breakthrough capacity per unit mass of adsorbent, wt%; Q_s is the saturation capacity per unit mass of adsorbent, wt%; v is the flow rate of fuel, cm^3/min; ρ is the measured fuel density, g/cm^3; c_i is the initial sulfur content, ppm; c_t is the sulfur content in the effluent fuel at time t, ppm; m is the mass of support packed in the column, g; x_i is the content of sulfur in fuel, wt%; t_b is the breakthrough time, min, and t_s is the saturation time when c_t/c_i=1, min.

Result and Discussion

Tests in model fuels

Desulfurization was operated at ambient pressure. The appropriate temperature is 70 °C for the removal of thiophen and 80 °C for benzothiophene as shown in Figures 1 and 2.

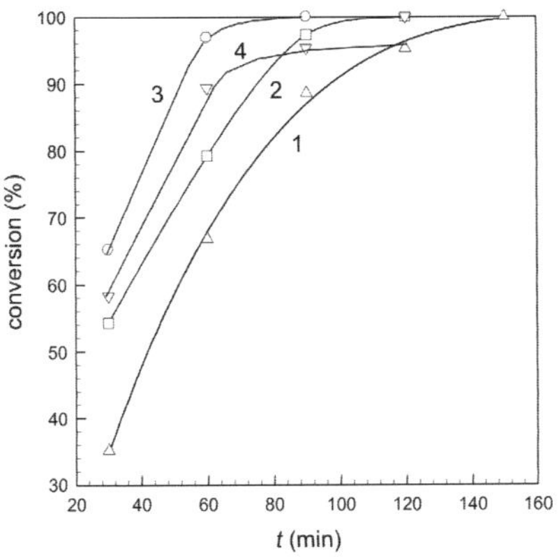

Fig. 1. Effect of temperature on thiophene conversion. 1: 50 °C; 2: 60 °C; 3: 70 °C; 4: 80 °C (Activated carbon as the carrier of H_2SO_4 and formaldehyde in ARF model fuel).

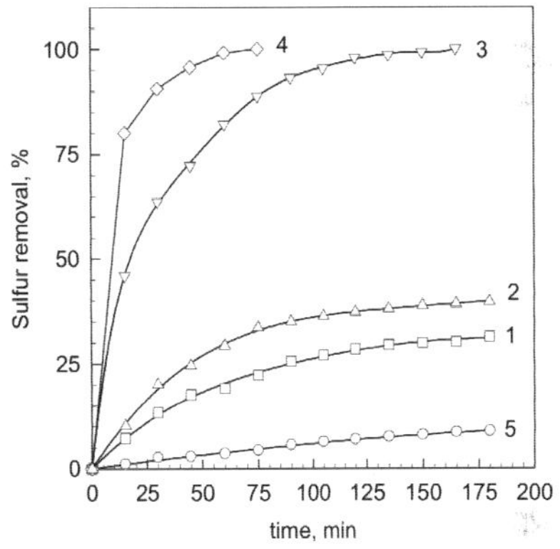

Fig. 2. Effect of temperature on the rate of removing 4,6-DMDBT contained in MXF fuel (catalyzed by PMA). 1: 50 °C; 2: 60 °C; 3: 70 °C; 4: 80 °C; 5: at 80 °C without loading formaldehyde.

The reaction rate increases following the increase of reagent (formaldehyde) loading amount from 20 to 40 wt % as shown in Fig. 3. However, too much quantity of catalyst (PMA) is not necessary as shown in Fig. 4. Larger quantity of reagent, longer period of operation for a given content of sulfur; therefore, larger specific surface area and/or pore volume is preferred in selecting adsorbent. Complete removal of 4,6-DMDBT from fuel may take longer time following the increase of sulfur content, but the content of 4,6-DMDBT does not much affect the desulfurization function as shown in Fig. 5. The function of desulfurization is valid for all types of fuels, though the aromatic fuel may retard the conversion rate to some extent, as shown in Fig. 6.

Tests in commercial fuels

The reaction/adsorption method of deep desulfurization was tested in commercial fuels bought from a gas station of Sinopec in Tianjin, China. The sulfur content in #93 gasoline is 480 ppm and that in #0 diesel is 1300 ppm and 1697 ppm respectively in two samples bought at different time. The sulfur content dropped to 202, 360 and 970 ppm, respectively in three samples. The remained sulfur compounds are not thiophenic, and most of them were removed through oxidation. The final sulfur content is below 15 ppm in all samples. The desulfured fuel was then treated with silica gel. The fuel quality of the clean fuel meets the National Standards of China, as shown in Table 1 for example.

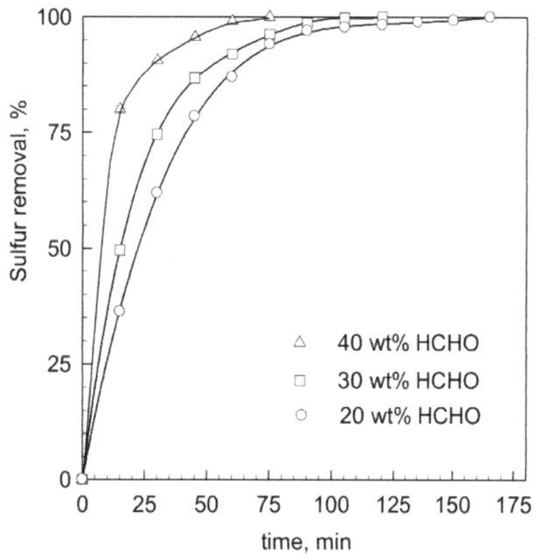

Fig. 3. Effect of loading quantity of formaldehyde on reaction rate (loading 70 wt % PMA at 80 °C).

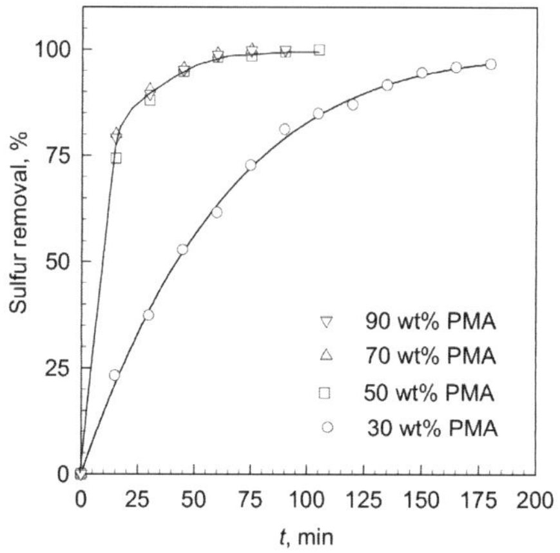

Fig. 4. Effect of loading quantity of PMA on reaction rate.

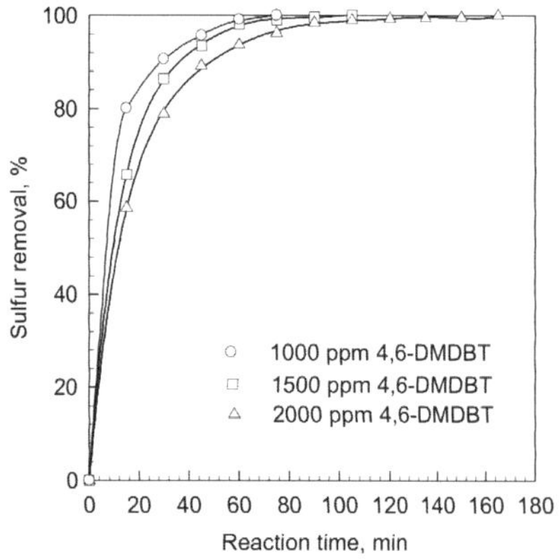

Fig. 5. Effect of 4,6-DMDBT content on reaction rate at 80 °C.

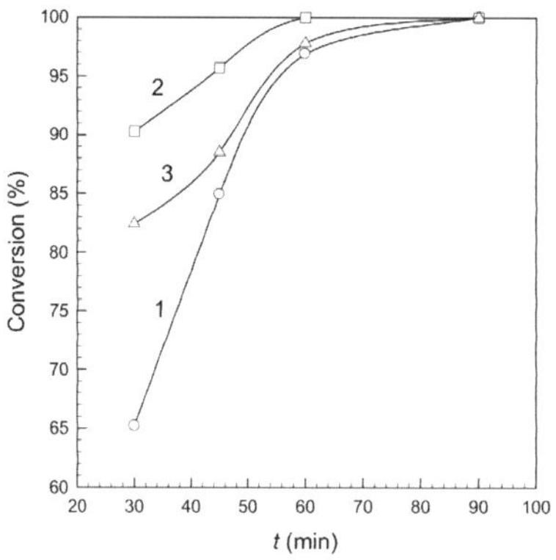

Fig. 6. Compatibility of the desulfurization method for different types of fuels. 1: ARF; 2: ALF; 3: MXF.

Regeneration

Two regeneration methods were tested. The first method is washing with alcohol, and most reaction products dissolved in it. However, little by-product was produced due to another reaction mechanism[32] and could not dissolve in alcohol. Therefore, the second method was used in the case that remarkable decrease in sulfur capacity was observed. The saturated carrier was heated at 400 °C in the air for 3 h and the by-product was decomposed and removed with a stream of inert gas. The PMA catalyst remained in regeneration, but formaldehyde needs to reload at room temperature. The sulfur capacity is basically recovered as shown in Fig. 7.

Table 1. Result of corrosion tests.

Test	Test standard	93# Gasoline	0# Diesel	Standard for gasoline (GB/T 17930-1999)	Standard for light diesel (GB 252-2000)
Copper corrosion	GB/T5096-85(91)	1a	1a	Less than Grade 1 at 50°C for 3 h	Less than Grade 1 at 50°C for 3 h
Acidity	GB/T258-77(88)	1.9	2.1	< 3	< 7
Soluble acid	GB/T259-88	None	None	None	None

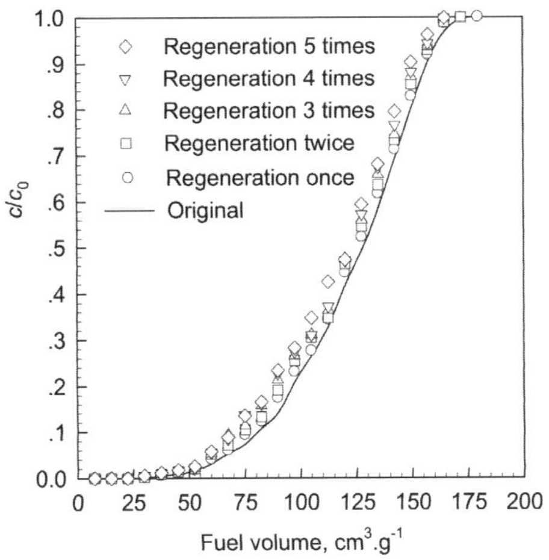

Fig. 7. Breakthrough curves of MXF fuel containing thiophene at 70 °C on regeneration.

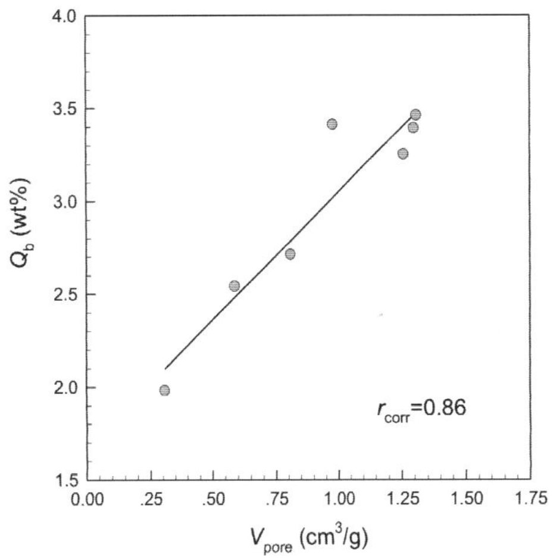

Fig. 8. Correlation of sulfur capacity with pore volume of carbon materials.

Preferred property of adsorbent

As the carrier of desulfurization reactions, the number, size and volume of adsorbent pores must affect the desulfurization performance. To test the dependence of sulfur capacity on porous structure, the breakthrough capacity was in turn correlated with the specific surface area, pore volume, and nominal pore size. Linear correlation coefficients obtained are 0.61, 0.89 and 0.50, respectively. It seems that pore volume exerts the largest effect and the correlation is shown in Fig. 8.

CONCLUSION

1. Reactions characteristic for sulfur species are effective for the deep desulfurization of transportation fuels provided the reaction is resided is an adsorbent.
2. The condensation reaction with formaldehyde and catalyzed by a strong acid completely removed thiophenes and benzothiophenes from transportation fuels. The remained sulfur species are removed by oxidation reaction.
3. Adsorbent can be regenerated with alcohol washing and heating in air at 400 ^{o}C. The washed out products are sulfur resin that may have other applications.
4. The proposed method is suitable for low sulfur content; therefore, is better to couple with the available HDS (hydrodesulfurization) technology.

Acknowledgement

Support of the National Natural Science Foundation of China (#20776099) is gratefully acknowledged.

References

1. S. Brunet, D. Mey, G. Pérot, C. Bouchy, F. Diehl (2005), “On the hydrodesulfurization of FCC gasoline: a review,” *Appl. Catal. A*. 278, 143-172.
2. E. Ito, J.A.R. van Veen (2006), “On novel processes for removing sulphur from refinery streams,” *Catal. Today*. 116, 446-460.
3. I.V. Babich, J.A. Moulijn (2003), “Science and technology of novel processes for deep desulfurization of oil refinery streams: a review,” *Fuel*, 82, 607-631.
4. C. Song (2003), “An overview of new approaches to deep desulfurization for ultra-clean gasoline, diesel fuel and jet fuel,” *Catal. Today*. 86, 211-263.
5. M. Breysse, G. Djega-Mariadassou, S. Pessayre, C. Geantet, M. Vrinat, G. Pérot, M. Lemaire (2003), “Deep desulfurization: reactions, catalysts and technological challenges,” *Catal. Today*. 84, 129-11138.
6. C. Schmitz, L. Datsevitch, A. Jess (2004), “Deep desulfurization of diesel oil: kinetic studies and process-improvement by the use of a two-phase reactor with pre-saturator,” *Chem. Eng. Sci*. 59, 2821-222829.
7. W.B. Wang, S.J. Wang, H.Y. Liu, Z.X. Wang (2007), “Desulfurization of gasoline by a new method of electrochemical catalytic oxidation,” *Fuel*, 86, 2747-2753.
8. O. Etemadi, T.F. Yen (2007), “Surface characterization of adsorbents in ultrasound-assisted oxidative desulfurization process of fossil fuels,” *J. Colloid. Interface Sci*. 313, 18-25.
9. D.S. Zhao, F.T. Li, E.P. Zhou, Z.M. Sun (2008), “Kinetics and Mechanism of the Photo-oxidation of Thiophene by O_2 Adsorbed on Molecular Sieves,” *Chem. Res. Chn. Univ*. 24, 96-100.
10. G. Yu, S. Lu, H. Chen, Z. Zhu (2005), “Oxidative Desulfurization of Diesel Fuels with Hydrogen Peroxide in the Presence of Activated Carbon and Formic Acid,” *Energy Fuels*. 19, 447-452.

11. M. Te, C. Fairbridge, Z. Ring (2001). "Oxidation reactivities of dibenzothiophenes in polyoxometalate/H_2O_2 and formic acid/H_2O_2 systems," *Appl. Catal. A*. 219, 267-280.
12. L. Yang, J. Li, X.D. Yuan, J. Shen, Y.T. Qi (2007), "One step non-hydrodesulfurization of fuel oil: Catalyzed oxidation adsorption desulfurization over HPWA-SBA-15," *J. Mol. Catal. A*. 262, 114-118.
13. F. Al-Shahrani, T.C. Xiao, S.A. Llewellyn, S. Barri, J. Zheng, H.H. Shi, G. Martinie, M.L.H. Green (2007), "Desulfurization of diesel via the H_2O_2 oxidation of aromatic sulfides to sulfones using a tungstate catalyst," *Appl. Catal. B*. 73, 311-316.
14. L.Y. Kong, G. Li, X.S. Wang (2004), "Mild oxidation of thiophene over TS-1/H_2O_2," *Catal. Today*. 341, 93-95.
15. P.S. Tam, J.R. Kittrell, J.W. Eldridge (1990), "Desulfurization of fuel oil by oxidation and extraction. I. Enhancement of extraction oil yield," *Ind. Eng. Chem. Res*. 29, 321-324.
16. X.L. Ma, A.N. Zhou, C.S. Song (2007), "A novel method for oxidative desulfurization of liquid hydrocarbon fuels based on catalytic oxidation using molecular oxygen coupled with selective adsorption," *Catal. Today*. 123, 276-284.
17. T.S. Jeyagowry, X. Huang, J. Dou, T.Y. Nah, R. Xu, P.K. Wong (2006), "A novel oxidative desulfurization process to remove refractory sulfur compounds from diesel fuel," *Appl. Catal. B*. 63, 85-93.
18. J. Eber, P. Wasserscheid, A. Jess (2004), "Deep Desulfurization of Oil Refinery Streams by Extraction with Ionic Liquids," *Green Chem*. 6, 316-322.
19. J. Gislason (2002), "Phillips sulfur-removal process nears commercialization," *Oil Gas J.* 99, 74.
20. R.L. Irvine (1998), US Patent 5,730,860, "Process for desulfurizing gasoline and hydrocarbon feedstocks," US Patent Issued on March 24.
21. A.J. Hernández-Maldonado, R.T. Yang (2004), "Desulfurization of Transportation Fuels by Adsorption," *Catal. Rev*. 46, 111-150.
22. R.T. Yang, A.J. Hernandez-Maldonado, F.H. Yang (2003), "Desulfurization of transportation fuels with zeolites under ambient conditions," *Science*, 301, 79-81.
23. C. Song, X. Ma (2003), "New design approaches to ultra-clean diesel fuels by deep desulfurization and deep dearomatization," *Appl. Catal. B*. 41, 207-238.
24. Y.W. Li, F.H. Yang, G.S. Qi, R.T. Yang (2006), "Effects of oxygenates and moisture on adsorptive desulfurization of liquid fuels with Cu(I)Y zeolite," *Catal. Today*. 116, 512-518.
25. Z.K. Zhang, X.L. Niu, S.L. Liu, X.X. Zhu, H.W. Yu, L.Y. Xu (2008), "The performance of HMCM-22 zeolite catalyst on the olefin alkylation thiophenic sulfur in gasoline," *Catal. Commun*. 9, 60-64.
26. T. Ohshiro, Y. Ishii, T. Matsubara, K. Ueda, Y. Izumi, K. Kino, K. Kirimura (2005), "Dibenzothiophene Desulfurizing Enzymes from Moderately Thermophilic Bacterium Bacillus Subtilis WU-S2B: Purification, Characterization and Overexpression," *J. Biosci. Bioeng*. 100, 266-273.
27. K. Kirimura, T. Furuya, Y. Nishii, Y. Ishii, K. Kino, S. Usami (2001), "Biodesulfurization of Dibenzothiophene and Its Derivatives through the Selective Cleavage of Carbon-Sulfur Bonds by a Moderately Thermophilic Bacterium Bacillus Subtilis WU-S2B," *J. Biosci. Bioeng*. 91, 262-266.
28. M. Graeser, E. Pippel, A. Greiner, J.H. Wendorff (2007), "Polymer Core-Shell Fibers with

Metal Nanoparticles as Nanoreactor for Catalysis," *Macromolecules*. 40, 6032-6039.

29. W. Dai, Y.P. Zhou, S.Q. Wang, W. Su, Y. Sun, L. Zhou (2008), "Desulfuriation of Transportation Fuels Targeting at Removal of Thiophene/Benzothiophene," *Fuel Proc Technol*. 89, 749-755.
30. S.Q. Wang, L. Zhou, L. Long, W. Dai, Y.P. Zhou (2008), "Thiophene Capture with Silica Gel Loading Formaldehyde and Hydrochloric Acid," *Ind. Eng. Chem. Res*. 47, 2356-2360.
31. S.Q. Wang, L. Zhou, W. Su, Y. Sun, Y.P. Zhou (2008) "Deep desulfurization of transportation fuels by characteristic reaction resided in adsorbents," Submitted and under review.
32. Y.L. Zhao, S.Q. He, C.D. Xu (1992), *Introduction to Chemistry of Heteropoly Acids* (in Chinese), Beijing, Higher Education Press, p. 144.

218c Role of the Soot Main Precursors in the Psds for Premixed Ethylene and Benzene Flames

Carlos A. Echavarria, Adel F. Sarofim, and **JoAnn Slama Lighty**. Chemical Engineering, University of Utah, Salt Lake City, UT 84112

Soot formation during combustion continues to be a major subject of experimental and theoretical study due to the impact of soot on human health and radiation forcing. However, soot formation is a complex process involving a great number of chemical and physical steps, and is still incompletely understood. Our previous study demonstrated the major differences in the evolution of the particle size distributions (PSDs), both measured and modeled, of soot in premixed benzene and ethylene flat flames. The most significant difference between these PSDs was the persistence of the nucleation mode with height above the burner (HAB) for ethylene flames compared to its rapid decline for benzene flames under similar temperature and C/O ratio conditions. Since conditions other than fuel changed between the flames, the question still remains as to whether the difference is accountable by the fuel change or by other flame conditions. In this study, we are measuring the concentration profiles of the soot main precursors acetylene, benzene and PAHs such as naphthalene, phenanthrene, pyrene and acenaphthalene in order to evaluate their role on the PSDs for these flames. Experimentally, samples are collected in a filter trap and cooled trap from the flames, the organic extractable material is analyzed using Gas Chromatography and the results are compared with different solutions standards prepared using benzene and light PAHs. The experimental concentrations profiles are compared with the model which employed calculations of gas phase species coupled with a discrete sectional approach for the gas-to-particle conversion. The model results showed that in the benzene flame the aromatics are consumed in the oxidation zone of the flame which might explain why the decrease in the nucleation mode in this flame. By contrast, in the ethylene flame, the acetylene and aromatics concentrations persist at even high HAB which is again consistent with the observed persistence of the nucleation mode at larger HAB

.

218d Application Vapor Dispersion Model of Natural Gas and Liquid Natural Gas (LNG) System to Estimate the Methane Gas Emission and Establish Correlation Between the Temperature and Concentration of the Dispersed Gas

Anisa Safitri and M. Sam Mannan. Mary Kay O'Connor Process Safety Center, Artie McFerrin Department of Chemical Engineering, Texas A&M University, 3122 TAMU, College Station, TX 77843-3122

The use of natural gas becomes more attractive due to its efficiency and cleaner combustion product and also as the oil price remains high, natural gas has been a potential energy source to displace it. However, according to U.S. EPA report on Global Anthropogenic Emission o f Greenhouse Gases non-CO2 2006, the methane emission from natural gas system accounts for 24% of the total methane emission to the atmosphere, and are produced from vented, fugitive leaks of the gas or accidental release of gas or liquid natural gas. Controlling methane gas emission is crucial not only it will reduce the economic loss, but also it reduces the greenhouse gas emission to the atmosphere. In addition, methane gas mixture is flammable in the range of 5% -15% of methane in the methane-air mixture and thus reducing the methane gas emission implied less hazardous material being released to the environment.

In this research the amount of methane emission will be estimated using vapor dispersion model and conservation of mass, energy and momentum will be established to predict the correlation of the downwind temperature and concentration of the dispersed methane gas. For natural gas system, Gaussian dispersion approach will be used to estimate the concentration and further will be modified based on the true parameters measured such as wind profile, humidity and solar radiation in the field test. The dense gas dispersion model will be used to estimate the emission from liquid natural gas release. Simulation of dense gas model using three-dimensional computational fluid dynamic (CFD) model will be presented.

218e Breakdown Mechanisms for Methyl-Esters

Kuang-Chuan Lin, Mechanical Engineering, University of Michigan, 2160 G.G. Brown, 2350 Hayward St, Ann Arbor, MI 48109-2125, Lam K. Huynh, Mechanical Engineering Department, University of Michigan, 2024 G.G. Brown, 2350 Hayward st., Ann Arbor, MI 48105, and Angela Violi, Mechanical Engineering, Chemical Engineering, Biomedical Engineering, University of Michigan, 2350 Hayward Str., 2150 G.G. Brown, Ann Arbor, MI 48109-2125.

Biofuels are liquid, solid, or gaseous fuels derived from renewable biological sources. Biomass can be burned directly for thermal energy or converted to other high-value energy sources including ethanol, biodiesel, methanol, hydrogen, or methane. Biodiesel is a biologically derived diesel fuel substitute created by chemically reacting vegetable oils or animal fats with an alcohol (methanol is the usual choice) to produce fatty acid methyl esters of the R-(C=O)-O-R' form (where R and R' are carbon chains of alkyl and alkenyl) with as many as 16-18 carbon atoms. Biodiesel is the name given to these esters when they're intended for use as fuel. Blends of up to 20% biodiesel (mixed with petroleum diesel fuels) can be used in nearly all diesel equipment and are compatible with most storage and distribution equipment. In this work we report a detailed analysis of the breakdown kinetic mechanisms for methyl-esters using theoretical approaches. Electronic structures and structure-related molecular properties of reactants, intermediates, products and transition states were explored at the BH&HLYP/cc-pVTZ level of theory. Rate constants for the main reaction pathways are computed for the temperature range of 300-2500 K.

218f Rules to Select Solvents for Scrubbing CO2 from Fossil Power Stations

Wolfgang Arlt and Liudmila Mokrushina. Chair of Separation Science and Technology, University of Erlangen, Egerlandstr. 3, Erlangen, Germany

The present amine-based technology scrubbs CO2 from flue gas by a chemical equilibrium reaction. Some disadvantages are obvious:

1. the recovery of the CO2-rich solution requires so much energy that the electrical output is reduced by 25-35 %

2. Amines are harmful to the environment and should be removed from the flue gas

3. Some amines form irreversible products

The focus of this lecture is to show how a thermodynamical analysis will lead to more economical solvents.

It turns out that the following properties are advantageous:

1. "no" vapour pressure

2. molecular weight around some few thousand Dalton

3. low viscosity

218g Clustering of Nano-Sized Carbon Particles Using Atomistic and Coarse-Grained Molecular Dynamics Simulations

Seung Hyun Chung, Mechanical Engineering, The University of Michigan, 2024 G.G. Brown, 2350 Hayward Str., Ann Arbor, MI 48109-2125 and Angela Violi, Mechanical Engineering, Chemical Engineering, Biomedical Engineering, University of Michigan, 2350 Hayward Str., 2150 G.G. Brown, Ann Arbor, MI 48109-2125.

The transition from gas-phase species to solid particles is probably the least understood part of the soot formation process. The rate of particle growth not only determines the final soot mass, but it also influences other soot properties such as morphology and composition. The capability to follow the nucleation and growth of nanoparticles in flames is crucial to arrive at a quantitative understanding of soot formation in flames. In this work, we use atomistic and systematic coarse-graining molecular dynamics simulations to study the clustering of nanoparticles of different size and chemical composition in high temperature regimes. The goal of these approaches is to overcome the limitations in time and length scales of the atomistic simulations achieving a simpler description of the effective interactions. Particles with carbon numbers ranging from 60 to 400 are coarse-grained and MD simulations are carried out at different temperatures. The clustering behavior function of temperature and size is analyzed in the temperature range 500-200K.

265a Catalytic Effects of Minerals Matters on Stable Surface Oxides and Oxidation Rate In the Early Stage of Coal Combustion

Shaolong Wan, Wei-Yin Chen, and Guang Shi. Chemical Engineering, University of Mississippi, Department of Chemical Engineering, Anderson Hall, University, MS 38677

Abstract

Reactivity of old chars has been characterized by temperature-programmed desorption (TPD) of surface oxides up to 1000 °C in the past. In a recent study, we discovered the existence of stable surface oxides on young chars that desorb only above 1100 °C.[1] After the wall interferences are minimized, we observed that oxygen from gas phase, organic portion of the coal and minerals in the coal have profoundly influences on the formation of stable surface oxides in the early stage of coal combustion.[2] In an attempt to isolate the effects of minerals, demineralized coals (DMC) are oxidized in O_2 with a contact time less than 1 second, and the amount and strength of stable surface oxides are characterized by TPD up to 1650 °C. Young chars derived from both demineralized lignite and bituminous coal show a low and flat TPD profiles over a wide temperature range signifying the minerals' catalytic activities in forming stable surface oxides, possibly in the adsorption step, for both coals. Indeed, the rates of oxidation of chars from both bituminous coal and lignite based on CO_2 and CO yields during oxidation show higher rates than their DMC counterparts. Moreover, the rate of oxidation seems to be qualitatively correlated to the amount of surface oxides. Similar to those from raw coals, combined oxygen balance and elemental analysis of chars from DMC suggests that that oxygen in the organic portion of the lignite activates oxygen turnover and carbon oxidation during its combustion; chars from neither raw nor demineralized bituminous coal possesses these properties. Minerals are known to play a significant catalytic role for combustion of lignite; it appears that after removing the minerals from bituminous coal both reaction rate and surface oxides released during TPD decrease too. XPS spectra of raw and demineralized bituminous coal and their char show peaks around 532.0 eV in O(1s) difference spectrum suggesting the existence of stable surface oxides.

Reference

1. Chen, W.Y., Wan, S., Shi, S., "Stable Surface Oxides on Chars and Impact of Reactor Materials at High Temperatures," *Energy & Fuels,* **2007**, 21, 778-792.

2. Chen, W.Y., Shi, G., Wan, S., "Stable Surface Oxides on Coal-Derived Chars," to be submitted to *Energy & Fuels,* **2008**.

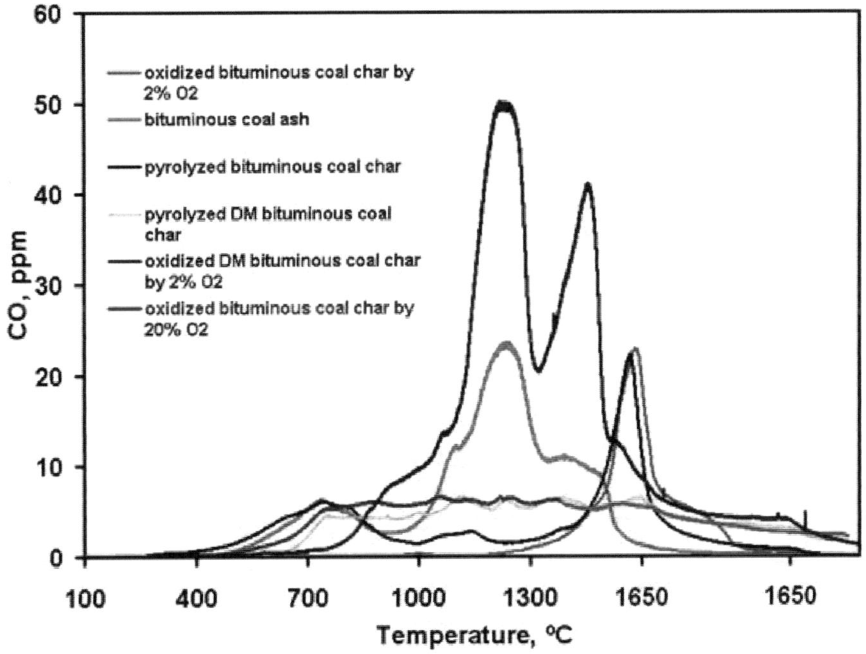

Figure 1. Temperature-program desoption (TPD) profiles of char oxidized by 2% O_2 and with 15% carbon burnout at 900 °C, char oxidized by 21% O_2 and with 75% carbon burnout, ash, pyrolyzed char, demineralized char oxidized by 2% O_2 and with 15% carbon burnout, and pyrolyzed demineralized char (all normalized based on 1 mg carbon in char). Peaks of the oxidized chars at 1250 °C suggest the existence of stable surface oxides that controls the rate of char oxidation in flame conditions. Peaks at higher temperatures (1450 and 1550 °C) represent interactions between the SiC tube and minerals in the coal. Chars derived from oxidation of DMC (lignite and bituminous) show low and flat TPD profiles. Moreover, DMC's also have lower oxidation rates. These observations suggest the possible catalytic role of minerals in the early stage oxidation of both lignite and bituminous coal.

Keywords: oxides, demineralized coal, oxidation, young char

* for presentation at the annual meeting of the American Institute of Chemical Engineers, Philadelphia, PA, November 16-21, 2008.

Wei-Yin Chen, contact author, Tel.: (662) 915-5651, Fax: (662) 915-7023, E-mail: cmchengs@olemiss.edu

265b Effect of Alkali Metals on Lignin Pyrolysis and Gasification

Vipul Kumar, Kristiina Iisa, Sujit Banerjee, and Wm. James Frederick Jr. School of Chemical and Biomolecular Engineering, Georgia Institute of Technology, 500 10th Street, NW, Atlanta, GA 30332

Lignin, the second most abundant naturally occurring material is a major constituent of biomass. Lignin is a waste produced during chemical pulping of wood and it is currently burned in recovery boilers to recover cooking chemicals and to produce steam. It is also produced during saccharification and fermentation of lignocellulosic biomass to ethanol. Lignin can be gasified to produce synthesis gas which can be converted to chemicals and fuels. This study focuses on the effect of alkali metals on pyrolysis and gasification of lignin. Three types of lignin were impregnated with varying amounts of alkali solutions at ambient conditions and were pyrolyzed and gasified at temperatures up to 800 °C. Alkali metals decreased the devolatilization rate and mass loss and increased the rate of gasification for all the lignins. Pyrolysis of all the lignins showed similar behavior until 600 °C but secondary reactions occurred at higher temperature for lignin with high ash content. These reactions were found to be due to interactions of char carbon with alkali metals salts. Gasification reactions are catalyzed by Na_2CO_3 and Na_2SO_4 with the rate increase being proportional to the amount of sodium added. The mechanism of alkali impact on pyrolysis and gasification is discussed.

Comparison and Validation of Gas Phase Reaction Schemes for Black Liquor Gasification Modeling

Per Carlsson*, Magnus Marklund, Henrik Wiinikka, Rikard Gebart

Energy Technology Centre (ETC), Box 726, SE-941 28, Piteå Sweden
*Corresponding author. Tel.:+46911232395; fax: +46911232399
E-mail address: per.carlsson@etcpitea.se

Introduction

Pressurized Entrained-flow High Temperature Black Liquor Gasification (PEHT-BLG) is a potential substitute or complement to the recovery boiler traditionally used for the recovery of chemicals and energy in black liquor in the Kraft pulping process. Black liquor consists of roughly 30 % moisture, 35 % inorganic pulping chemicals and 35 % combustible material (i.e. lignin). The PEHT-BLG technology can give an increase in total energy efficiency of the mill and provide new products with high added value, such as green motor fuels. The main parts of the recovery unit in the process are (see figure 1); a slagging refractory lined entrained-flow gasification reactor, with a gas assisted burner nozzle producing small black liquor droplets, used for direct gasification of the black liquor at about 1000 °C to produce a 'raw' syngas and a liquid smelt containing mainly Na_2CO_3 and Na_2S; a quench cooler beneath the reactor where the product gas and smelt are separated and the smelt is dissolved in water forming green liquor; a counter current condenser (CCC) that cools the syngas and condenses water vapor and any volatile and tar species that may be present. The heat recovered from the gas condensation is used to generate low/medium pressure steam that can be used in the pulp and paper process. Furthermore, the chemicals in the green liquor are recovered as cooking chemicals in the downstream processing.

Due to lack of demonstration of long term operation of the technology, a development (pilot) plant for PEHT-BLG (named DP-1) with a capacity of 20 tones dry solids/24h is in operation by the technology vendor Chemrec AB at the Energy Technology Centre in Piteå, Sweden [5]. An important tool for reduction of the technical risk associated with scale up of new technology is a comprehensive CFD model for the PEHT-BLG reactor that has been developed by Marklund [8]. The model includes drying, pyrolysis, char gasification and smelt formation of black liquor droplets as well as a simplified gas phase reaction mechanism.

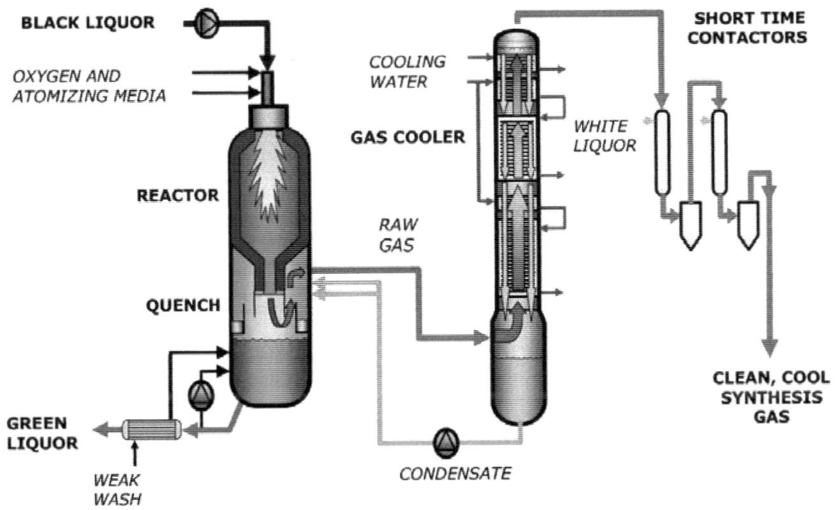

Figure 1. Schematic drawing of the PEHT-BLG process (Courtesy of Chemrec AB)

Marklund et. al [8] has made an initial validation of the model against the outlet gas composition after the Counter Current Condenser (CCC), see figure 1. The model predicted a CO / CO_2 ratio that was approximately 50% higher compared to the measurements. However, it is possible that the well known water-gas shift reaction is active in the quench and this could explain that the experimentally determined gas composition after the CCC differs from the computational results at the outlet from the hot zone.

Recently, in-situ measurements have been performed in the DP-1 reactor and a further validation of the model has been made possible. The measurements have been performed by sampling gas with a water-cooled suction probe from the lower part of the hot zone, followed by offline gas analyses. The details regarding the probe sampling equipment and procedure will be described elsewhere.

The first measurements in the hot reactor showed that; reactions in the quench vessel and following CCC did not change the gas composition significantly. The over-prediction of the CO / CO_2 ratio was therefore assumed to be caused by a mechanism not included in Marklund's model [9]. The model relies on a four step simplified gas phase reaction scheme for hydrocarbons, proposed by Jones and Lindstedt [4]. This reaction scheme does not include oxidation of CO by O_2 and was believed to be the main contributor to the over-predicted CO / CO_2 ratio.

The present paper investigates the difference between Marklund's model [9], which uses the reaction scheme by Jones and Lindstedt for the gas phase, and a modified model with the additional $CO + O_2$ reaction added to the Jones - Lindstedt reaction mechanism as proposed by several other authors [2, 11, 13]. The simulation results are compared against measurements obtained by the gas sampling probe in the DP-1 reactor.

Geometry and Mathematical Model

The DP-1 reactor is an axi-symmetric entrained-flow reactor with the spray burner centrally placed at the reactor top, see figure 1. The reactor dimensions are 2.3 m in height and 0.6 m in inner diameter. Based on the rotational symmetry, the reactor geometry was modeled as a 2D slice using periodic azimuthal boundary conditions. Heat losses were neglected in this study and the refractory lined reactor wall was modeled as adiabatic. However, rough estimates have shown that the heat loss through the reactor wall is of the order of 100 kW or about 3% of the total thermal fuel power.

Marklund [9] implemented his model in the commercially available finite volume method based CFD program, ANSYS CFX 4. Since then, the model has been implemented in ANSYS CFX 11 by the present authors to take advantage of the improved numeric schemes in the later version of the code. The implementation of the current model into CFX 11 has been made with particle user Fortran routines in a similar manner as in [8] and should in theory yield the same results. The details of the model are described thoroughly by Marklund [9] and will therefore only be described in brief below.

The burner was modeled as a simplified spray burner with concentrical annular inlets where oxygen and discrete black liquor droplets enter the gasifier at a prescribed angle and velocity. The dispersed black liquor particles were modeled using the Euler – Lagrange formulation [1]. In the present paper the black liquor spray was represented by 1003 discrete particles having a fitted Rosin Rammler size distribution. The droplet size distribution and flow velocity are consistent with data from nozzle experiments measured by phase Doppler anemometry.

As the black liquor droplets pass through the reactor they undergo four main steps of conversion: drying, devolatilization, char gasification and finally smelt formation. During the different conversion stages there is a large mass transfer from the particles to the gas phase. During drying, water is vaporized, during devolatilization gaseous species (CO, CO_2, CH_4, H_2 and H_2S) are released and when the remaining char is gasified CO, CO_2 and H_2 are released.

In the model by Marklund [9] the simplified reaction scheme described by Jones and Lindstedt was used for the gas phase reactions, see reaction R1 to R4. Only CH_4 was considered in the model, hence n=1.

$$C_nH_{2n+2} + \frac{n}{2}O_2 \rightarrow nCO + (n+1)H_2 \qquad \text{R1}$$

$$C_nH_{2n+2} + nH_2O \rightarrow nCO + (2n+1)H_2 \qquad \text{R2}$$

$$H_2 + \frac{1}{2}O_2 \rightarrow H_2O \qquad \text{R3}$$

$$CO + H_2O \leftrightarrow CO_2 + H_2 \qquad \text{R4}$$

Several authors [2, 11, 13] has included an additional reaction (R5) to the gas phase reaction scheme above when modeling gasification and achieved acceptable results.

$$CO + \frac{1}{2}O_2 \rightarrow CO_2 \qquad \text{R5}$$

In the present paper the simplified reaction scheme by Jones and Lindstedt (reactions R1-R4) was implemented. The influence of reaction R5 was also implemented together with reaction R1 to R4 to see the response in flame temperature and outlet gas composition.

Due to numerical difficulties, kinetics could not be included for all reactions. Reaction R1, R2 and R5 was modeled using only the Eddy Dissipation Model (EDM) [7]. Reaction R3 and R4 was modeled using a combination of Finite Rate Chemistry (FRC) and EDM using the kinetics from Jones and Lindstedt. The EDM-FRC model calculates the turbulent and kinetic reaction rate and uses the minimum of the two. The simplified reaction scheme described by reactions R1 to R4 will be referred to as case 1, and reactions R1 to R5 to case 2.

To model turbulence the k-ε model with standard wall functions [12] was used. The radiative heat transfer was modeled using the Discrete Transfer (DT) radiation model by Lockwood and Shah [6] treating the wall as optically smooth with a radiative emissivity of 0.5. The absorption coefficient for the gas was calculated as the mass weight average of the participating species. Particles that hit the wall were assumed to lose 50% of their momentum in the perpendicular direction and none in the parallel direction in order to simulate a wall film flow. Initial calculations with deterministic particle trajectories suffered from poor convergence and a large amount of the particles became trapped in the recirculation zone. By introducing a model for turbulent dispersion as suggested by Gosman and Ioannides [3] this phenomenon was reduced.

In the current model the proximate analysis done by Marklund [9] was used with some modification based on more recent elementary and heating value data. The proximate analysis results are presented in table 1, the composition of the volatile matter in table 2 and initial smelt composition in table 3.

Table 1. Proximate analysis result, the weight fractions are consistent with elemental analysis.

Proximate matter	%wt
Moisture	30.00
Volatile matter	27.67
Char in smelt	9.38
Smelt (ash)	32.95

Table 2. Volatile matter composition as released during devolatilization.

	Specie	%wt
Volatile matter	H_2S	7.66
	CO	53.13
	CO_2	10.44
	H_2	2.26
	CH_4	26.51

Table 3. Initial smelt composition present in the virgin black liquor.

	Specie	%wt
Smelt	Na_2SO_4	13.13
	Na_2S	1.03
	Na_2CO_3	85.84

Results

Experimental

Four measurements were made in the DP-1 reactor with the gas sampling probe placed close to the outlet but still inside the hot zone. The oxygen / dry black liquor mass flow ratio (kg/kg) and pressure were held constant at 0.42 and 27 bar respectively. The synthesis gas samples that were withdrawn from the reactor were analyzed for: CO, CO_2, CH_4, H_2 and H_2S using a Varian CP-3800 gas chromatograph. The resulting dry average gas composition from the measurements in DP-1 is presented in table 4 together with the average LHV.

Table 4. Average molar gas composition ± one standard deviation, CO / CO_2 ratio and average LHV. The measurements were made with the probe in the DP-1 reactor at 0.42 oxygen / dry black liquor ratio (kg/kg) and 27 bar pressure.

Specie	Gas composition (%mole)
CO	27.6 ± 0.83
CO_2	33.7 ± 0.46
CH_4	1.2 ± 0.05
H_2	36.2 ± 0.71
H_2S	1.3 ± 0.05
CO / CO_2	0.82
LHV	7.62 MJ/kg

During the measurements all relevant process parameters, e.g. mass flow rates, temperatures and operating pressure, were logged using the process monitoring system available in DP-1.

Computational

The computed gas composition was sampled at a point that corresponds to the position of the hot probe tip used in the measurements. The results are presented in table 5 with the H_2O excluded so a comparison can be made between the calculated and measured values.

Table 5. Molar gas composition, CO / CO_2 ratio and LHV from the PEHT-BLG-CFD model with reaction R1-R4 implemented (left) and with extended reaction scheme R1-R5 (right).

Specie	Case 1 R1-R4 Outlet gas composition (%mole)	Case 2 R1-R5 Outlet gas composition (%mole)
CO	33.11	33.49
CO_2	26.01	26.36
CH_4	0.00	0.00
H_2	38.87	38.12
H_2S	2.01	2.03
CO / CO_2	1.27	1.27
LHV	7.85 MJ/kg	7.77 MJ/kg

The volume average gas temperature was calculated over the complete modeled reactor. The difference between the volume average gas temperature, the sampling point gas temperature and the reactor outlet gas temperature was minimal for each case, see table 6. The peak gas temperature was significantly higher in case 2 compared to case 1.

Table 6. Calculated gas temperatures from the PEHT-BLG-CFD model with reaction R1-R4 implemented (left) and with extended reaction scheme R1-R5 (right).

	Case 1 R1-R4 Temperature (K)	Case 2 R1-R5 Temperature (K)
Peak gas temperature	2866	3576
Volume average gas temperature	1322	1366
Sampling point gas temperature	1331	1372
Reactor outlet gas temperature	1330	1372

Discussion

The peak temperatures in the simulations are noticeably high. However, an estimate of the peak temperature, assuming that hot CO is recirculated from the lower part to the top of the gasifier where it reacts with O_2 from the burner nozzle, shows that the temperatures are thermodynamically possible. Figure 2 shows a vector plot from the simulations and the flow pattern is consistent with the assumption that hot CO can be convected to the top of the reactor. Assuming that CO is entrained into the shear layer created by the oxygen jet that is used for atomization of the black liquor and the resulting CO / O_2 mixture has a temperature of about 900K results in an adiabatic flame temperature of about 3570K at 27 bar which is close to the computed values. The calculation of the adiabatic flame temperature was done with the chemical equilibrium program Gaseq [10] using thermodynamic data from the database in this program.

The volume average gas temperature and outlet gas temperature are much lower and are reasonably close to the temperatures measured with wall mounted and shielded thermocouples in the DP1 gasifier. It is well known that measurements in high temperature environments are difficult due to the large radiative heat flux to the temperature sensor. It is therefore likely that the experimental temperatures also differ from the true gas temperature and the validation of the computed gas temperature is therefore not possible based on the current experimental set-up.

Validation of the temperatures and especially the peak temperature are extremely difficult due to the highly corrosive atmosphere inside the reactor. However, more detailed temperature measurements with a different sensor technology are planned.

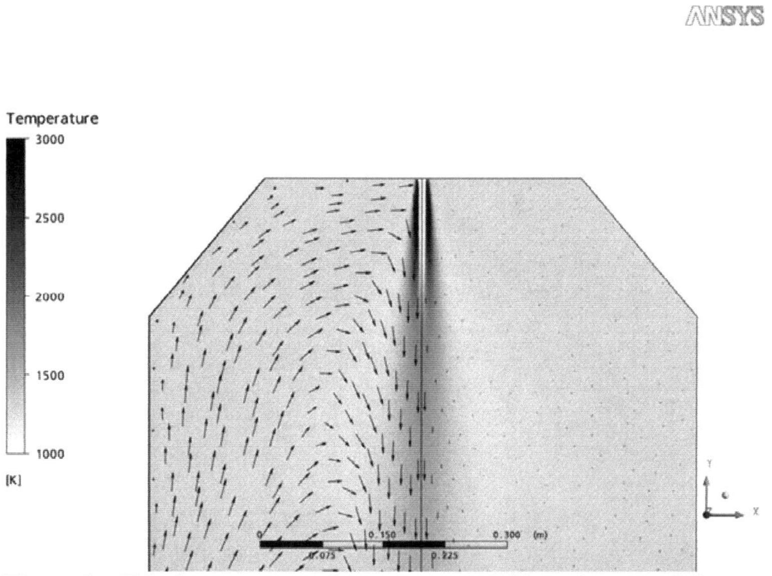

Figure 2. Velocity vectors and temperature field at the top of the reactor from the PEHT-BLG-CFD model.

Since wall heat losses are neglected in the model it can be assumed that the simulated gas temperatures are higher than in the DP-1 reactor. The elevated temperature will shift the chemical equilibrium of the water gas shift reaction (R4) towards the left i.e. a larger concentration of CO. Implementation of wall heat losses in the model would lower the gas temperature which would yield a gas composition closer to the one measured in the DP-1 reactor.

H_2S is currently implemented as an inert gas species in the model; the concentration of H_2S in the outlet gas is therefore determined already when specifying devolatilization species and mass fractions.

CH_4 is not present in the outlet gas composition from either of the models. This is consistent with thermodynamic equilibrium but differs from the experiments where the CH4 concentration is about 1.2%. The most likely reason for this is that tars are not included in the models. In the real gasifier, lignin is decomposed into heavy volatile compounds (tars) which in turn are decomposed further until equilibrium is achieved. It is possible that the kinetics of this chain of reactions is such that a significant fraction of CH_4 is still unreacted when the gas exits the gasifier. In the models; the CH_4 released during devolatilization will immediately be oxidized to CO and H_2 (reaction R1 and R2) limited only by the rate of mixing. If tars were included into the model, it would be possible to implement mechanisms for the thermal decomposition of tars to CH_4. However, validation of the mechanisms are difficult and the implementation would have to be made more or less *ad hoc*.

Conclusions

It has been showed that by implementing reaction 5 into the PEHT-BLG CFD model developed by Marklund [8] the local flame temperature was increased significantly. The effect on volume average and outlet gas temperature was minimal. It has also been showed that reaction 5 has very little effect on outlet gas composition when used with the Jones – Lindstedt simplified reaction scheme.

The peak temperatures in both models are very high while the volume average and outlet temperatures are consistent with measurements. However, it was shown that the computed temperatures are thermodynamically possible if one assumes that hot CO is entrained into the shear layer of the oxygen jet that is used for atomization of black liquor.

For the improvement of the model, heat losses through the reactor wall should be investigated further. It is also of interest to investigate chemical reaction models that can predict the experimentally determined levels of CH_4 at the outlet of the gasifier,

Acknowledgments

This work has been made possible by the funding from the BLG II program. This program is sponsored by the Swedish Energy Agency, Mistra, Kappa Kraftliner AB, SCA Packaging AB, Södra's Research Foundation, Sveaskog AB, Chemrec AB and the County Administrative Board of Norrbotten. The authors would also like to thank Carola Grönberg and Marcus Lidman for the help when conducting the experiments in DP-1.

References

1. Crowe C., Sommerfeld M. and Tsuji Y. (1998). *Multiphase flows with droplets and particles* Boca Raton: CRC press

2. Fletcher D.F., Haynes B.S., Christo F.C. and Joseph S.D. (1999). A CFD cased combustion model of an entrained flow biomass gasifier. *Applied Mathematical Modelling* **24**, 165-182

3. Gosman A.D and Ioannides E. (1981). Aspects of computer simulation of liquid fuelled combustors. AIAA paper No. 81-0323

4. Jones W.P. and Lindstedt R.P. (1988). Global Reaction Schemes for Hydrocarbon Combustion. *Combustion and Flame* **73**, 233-249

5. Lindblom, M. and Landälv, I., "Status of the Swedish National Black liquor Gasification (BLG) Develoment Program", TAPPI EPE Conference 2006. TAPPI, Atlanta (2006).

6. Lockwood F.C. and Shah N.G. (1981). A new radiation solution method for incorporation in general combustion prediction procedures. *Proc. Combust. Inst.* **18** pp. 1405-14.

7. Magnussen B.F. and Hjertager B.H. (1976). On mathematical modelling of turbulent combustion with special emphasis on soot formation and combustion. *Proc. Combust. Inst.* **16**, pp. 719-27

8. Marklund M., Gebart R., and Tegman R. (2007) Experiments and reactor model predictions in high temperature black liquor gasification. *Proc. International Chemical Recovery Conference*, May 29-June 1 2007, Quebec, Canada, pp. 147-154.

9. Marklund M., Tegman R., and Gebart R. (2008) A self-consistent CFD-model for high temperature pressurised black liquor gasification. *IFRF Combustion* Journal no. 200801, ISSN 1562-479X

10. Morley C. (2005) Gaseq. Chemical equilibria in perfect gases. Version 0.79. Available at http://www.gaseq.co.uk/ (September 22, 2008).

11. Watanabe H. and Otaka M. (2006). Numerical simulation of coal gasification in entrained flow coal gasifier. *Fuel* **85,** 1935-1943

12. Wilcox D.C. (1993) *Turbulence modelling for CFD*, Glendale: Griffin printing.

13. Yang W., Ponzio A. and Lucas C, Blasiak W. (2005). Performance analysis of a fixed-bed biomass gasifier using high-temperature air. *Fuel Processing Technology* **87,** pp. 235-245

Scrap Tires in Waste-to-Energy Plants: Treatment and Emission Control

Eilhann Kwon & Marco J. Castaldi
Department of Earth and Environmental Engineering [HKSM]
Columbia University in the City of New York, NY, 10027

The conversion of Municipal Solid Waste (MSW) to energy in Waste-to-Energy (WTE) facilities has been recognized as one of the means of effective waste management. The process concentrates the combustible components of MSW, and results in a fuel such as a waste tire with higher value. A series of Thermo-Gravimetric Analysis (TGA) coupled to Gas Chromatography/Mass Spectroscopy (GC/MS) has been done under various heating rates and atmospheres in order to understand and investigate the decomposition of waste tires via combustion and gasification/pyrolysis. The major objective of this work is to develop a mechanistic understanding of waste tire decomposition from starting constituents to final products.

One of the most significant findings was the observation of a plateau in the reaction at high temperature with oxygen that divided the thermal degradation into a primary and secondary reaction mechanism. The primary mechanism induced devolatilization while the secondary reaction mechanism induced a combustion reaction whose rate was proportional to the oxygen concentration. Direct evidence of this was the identification of epoxide chemical species using GC/MS analysis of produced gases.

Chemical releases from the thermal degradation of SBR, IR, and a waste tire have been measured with a GC/MS at heating rate 10C°/min and 20C°/min in various atmospheres. Chemical analysis of a waste tire using a GC/MS ensures that attribution of the thermal degradation products arising from SBR and IR is significant and measurably distinguishable. A significant amount of benzene derivatives were observed leading to conditions that was favorable for generating Polycyclic Aromatic Hydrocarbons (PAHs). Thus, benzene derivatives were investigated intensively along with their structural isomers, which lead to a mechanistic understanding of PAH formation.

Finally an attempt to approach more realistic test condition was made using a flow-through reactor for pyrolysis of real waste tire. Significant Volatile Organic Carbons (VOCs) including benzene derivatives, PAHs, and Hetero N-containing PAHs were observed. This study provides new data for the overall development and validation of detailed reaction mechanisms that can describe the thermal degradation of waste tire.

265e Pilot-Scale Comparison of Steam Gasification with Herbaceous and Woody Feedstocks

Katherine Gaston, Calvin Feik, Mark Nimlos, Daniel Carpenter, and Stephen Phillips. National Bioenergy Center, National Renewable Energy Laboratory, 1617 Cole Blvd., M/S 3311, Golden, CO 80401

Production of renewable transportation fuels from biomass has become an important priority to reduce the nation's dependence upon imported oil and minimize the impact upon the environment. However, a sustainable biofuels economy will require conversion processes that utilize the lignocellulosic components of plants so as not to compete with food production. To this end, the National Renewable Energy Laboratory's gasification research focuses on developing processes that use a variety of feedstocks, such as agricultural residues, waste from forest thinning, and energy crops. Biomass gasification combined with catalytic synthesis has the potential to meet these criteria. This approach can potentially produce alcohols or Fischer-Tropsch fuels from a variety of biomass sources.

Key to the economic viability of gasification/catalytic synthesis processes is the production of a syngas with low concentrations of products which interfere with the catalytic step, such as methane, tars and sulfur compounds. Developing an understanding of the formation of these species as a function of feedstock is critical for evaluating and optimizing gasification. In this study, steam gasification was conducted on a ten to twenty kilogram per hour (pilot) scale of mixed hardwoods, corn stover, switchgrass and wheat straw. Process conditions such as steam-to-biomass ratio and temperature were investigated and their effects upon product formation. Real-time measurements of gaseous products were gathered, including sulfur compounds. The chemical composition of the tars formed was monitored in real time using a Molecular Beam Mass Spectrometer (MBMS). The process and unit operations, operational lessons-learned, and results from the experiments as they relate to the formation of undesirable products will be discussed.

265f A Simple Gasifier Model That Runs in Aspen Dynamics

William L. Luyben and Patrick J. Robinson. Chemical Engineering, Lehigh University, Bethlehem, PA 18015

Gasification has been used in industry on a relatively limited scale for many years, but it is immerging as the premier unit operation in the energy and chemical industries. The switch from expensive and insecure petroleum to solid hydrocarbon sources (coal and biomass) is occurring due to the vast amount of domestic solid resources, national security and global warming issues. Gasification (or partial oxidation) is a vital component of "clean coal" technology. Sulfur and nitrogen emissions can be reduced, overall energy efficiency is increased and carbon dioxide recovery and sequestration are facilitated. Gasification units in an electric power generation plant produce a fuel gas for driving combustion turbines. Gasification units in a chemical plant generate synthesis gas, which can be used to produce a wide spectrum of chemical products. Future plants are predicted to be hybrid power/chemical plants with gasification as the key unit operation.

The widely used process simulator Aspen Plus offers a rigorous gasifier model that handles solids, so steady-state design and optimization studies of processes with gasifiers can be undertaken. However, at the present time these models cannot be exported into Aspen Dynamics. Dynamic simulations are essential for the development of stable and agile plantwide control structures for energy and chemical processes.

This paper presents a simple approximate method for achieving the objective of having a gasifier model that can be exported into Aspen Dynamics. The basic idea is to use a high molecular weight hydrocarbon that is present in the Aspen library as a pseudo fuel. This component should have the same 1:1 hydrogen-to-carbon ratio that is found in coal and biomass. For many plantwide dynamic studies, a rigorous high-fidelity dynamic model of the gasifier is not needed. The proposed approximate model captures the essential macro-scale thermal, flow, composition and pressure dynamics.

552a Oxy-Coal Combustion: Effects of PO2 on Coal Jet Stability In O2/CO2 Environments

Jingwei Zhang, Eric G. Eddings, and Jost O.L. Wendt. Department of Chemical Engineering & Institute for Clean and Secure Energy, University of Utah, 3290 MEB, 50 S. Central Campus Dr., Salt Lake City, UT 84112

The purpose of this research is to understand and predict the effects of the near burner zone environment on ignition characteristics of a turbulent axial pulverized coal jet. In oxy-coal combustion this environment consists of O2 and CO2 (instead of N2). Ultimately it is planned to provide data for subsequent simulation validation studies that can be used to predict how air fired combustors may be retrofitted to oxy-coal.

Specifically the objective is to explore effects of variations in the partial pressure of O2 and CO2 on coal jet ignition and flame stability. During the oxy coal jet ignition process, the partial pressure of oxygen, PO2, in the transporting fluid, comprised one additional degree of freedom in controlling flame aerodynamics, and enhancing coal ignition, flame attachment/detachment and modifying pollutant formation. The experimental investigations are carried on in a 100 kW laboratory furnace outfitted with an axial burner and arrays of electrically-heated panels in the burner zone in order to control variations in near-burner heat loss. The furnace also has visual access to the burner zone through quartz windows. Sufficient supply of O2 and fresh CO2 are supplied through special utility infrastructure in the laboratory. A special CMOS sensor based camera is used to capture Type 0 axial turbulent diffusion flame shape for the use of statistical studies of flame length at different operational parameters, with special emphasis on systematic variations of partial oxygen pressure in both transport and secondary oxidant streams. Our experimental data show how flame attachment increased with PO2 in either primary or secondary oxidant streams. These variables were also shown to have a first-order influence on fuel NOx. Future work will use these data to validate CFD simulations of axial coal jet ignition in turbulent flows..

Keywords: Oxy-coal Combustion, flame stability, oxygen partial pressure (PO2), flame length, near burner phenomena, coal jet, type 0 axial turbulent flames

552b Role of Recycle on Emissions from Oxy-Coal Combustion

S. Michael Daukoru, **Ying Li**, and Pratim Biswas. Energy, Environmental and Chemical Engineering, Washington University in St. Louis, 1 Brookings Drive, St. Louis, MO 63130

Oxy-coal combustion is being touted as a potential means for increasing concentration of carbon dioxide in exhaust gases to help in potential remediation and sequestration. The impact of this modality of combustion on other emissions such as fine particles, trace metals and mercury has been examined in both drop tube and self sustaining coal combustors (Suriyawong et al., 2006). Most of these studies have been conducted by combustion of the coal in a mixture of oxygen and carbon dioxide. However, in a full scale operating scenario, the flue gases will be recycled with only the necessary amount of oxygen added at the inlet with the coal.

In this paper, we will compare the effect of flue gas recycle on fine particle formation and other emissions, to that of the controlled studies performed earlier. Due to the difference in the temperature-time histories and the composition of the flue gases, there are expected to be differences in the combustion characteristics. First, the sub-oxide formation pathways are expected to result in different vaporization rates due to the different time-temperature relationships. It is expected that the more volatile elements such as mercury will not show any differences, however, the chemistry in the gas phase maybe different due to the potential for oxidation to increase. Clearly, the performance of added sorbents – both micrometer and nanostructured, will be different – and the results will be reported. Strategies for effective modalities for oxy-coal combustion to minimize emissions of fine particles and other toxic elements such as mercury will be proposed.

Suriyawong, A., Gamble M., Lee M.H., Axelbaum R. and Biswas P., "Submicrometer particle formation and mercury speciation under O2 - CO2 Coal Combustion", Energy and Fuels, 20 (6), 2357-2363, 2006.

552c Modeling, Construction and Operation of An Integrated Pollutant Removal Process at An Oxy-Combustion Test Facility

Stephen J. Gerdemann, Thomas Ochs, Danylo Oryshchyn, and Cathy Summers. National Energy Technology Laboratory, U.S. Department of Energy, 1450 Queen Avenue, SW, Albany, OR 973212152

Oxy-fuel combustion of coal uses oxygen rather than air as an oxidant resulting in a flue gas containing primarily CO2 and water. Using NETL's Integrated Pollutant Removal (IPR) process the flue gas can be cooled condensing and removing the water, and then compressed to produce a CO2 stream suitable for sequestration. The cost of energy for producing the oxygen and running the compressors reduces the efficiency a sub critical pulverized coal power plant approximately 5 percentage points from 35 to 30%. By careful design it is possible to recover some of this energy improving the efficiency to 33%. Models of an oxy-combustion power plant with IPR show how material and energy flow through the process and where energy can be recovered. In conjunction with the modeling effort an instrumented IPR process is being constructed at the Jupiter Oxygen oxy-combustion burner test facility. These experiments will measure gas composition as well as temperatures, pressures and flow rates throughout the process. These results will then be used to further refine the model of the process. The results will also be used in new computational fluid dynamics models as well as in economic models.

552e Oxygen Sorbents for Oxy-Fuel Combustion

Gokhan O. Alptekin[1], Ambalavanan Jayaraman[1], Margarita Dubovik[2], and Lauren Brickner[1]. (1) TDA Research, Inc, 12345 W. 52nd. Avenue, Wheat Ridge, CO 80033, (2) TDA Research Inc., 12345 W. 52nd Ave, Wheat Ridge, CO 80033

Oxy-fuel combustion has the potential to dramatically decrease the environmental impact of producing electrical power, especially because it can be retrofit to existing pulverized coal power plants. In this technology, oxygen combustion replaces the traditional combustion environment of coal in air. This results in an off gas of steam and pure CO_2, which can be easily sequestered rather than released into the atmosphere where it contributes to the green house effect. However, the technology is not yet used in spite of its promise, primarily because pure oxygen is currently too expensive. TDA Research, Inc. (TDA) is developing a low cost, high capacity sorbent for air separation to supply low cost oxygen for oxy-fuel combustion. The regenerable sorbent can remove oxygen from air with very high capacity, and the sorbent maintains its capacity for over 200 absorption/regeneration cycles under representative conditions. TDA's oxygen sorbents also would find application in IGCC plants. Using oxygen instead of air in the IGCC process will remove excess nitrogen from the gasifier and there by increase efficiency, reduce NOx emissions and improve gas purity. Removing nitrogen from the fuel gas is particularly critical if the synthesis gas is used for the production of chemicals instead of electricity.

552f Oxycoal Firing in Circulating Fluidized Bed Boilers

Eric G. Eddings[1], Astrid E. Sánchez[2], Liyong Wang[1], and Fanor Mondragón[2]. (1) Dept. of Chemical Engineering, University of Utah, 50 So. Central Campus Drive, 3290 MEB, Salt Lake City, UT 84112-9203, (2) Institute of Chemistry, University of Antioquia, A.A. 1226, Medellín, Colombia

This research program has two primary objectives: 1) to obtain fundamental rate and fuel nitrogen conversion information for oxycoal-fired circulating fluidized bed (CFB) boilers; and 2) to obtain model validation data from pilot-scale operation of an oxycoal-fired CFB.

Fundamental Studies

The fundamental studies are being carried out utilizing both bench-scale experiments and computational chemistry techniques. Previous studies have shown that recirculated CO2 not only reduces flame temperature, but also has a chemical effect that is reflected by increasing the CO formation that modifies oxidative routes, especially the formation and reduction of NO [1]. Another study [2] explored the effect of CO2 on NO emissions, using O2/CO2 mixtures and recirculated flue gas, as compared to conventional air combustion, and found reduced NO levels for the O2/CO2 mixtures. Although this reduction could be attributed to the reduction of Thermal NO (by removal of N2 as well as the use of flue gas recycle for temperature control), it has also been observed [3] that combustion in O2/CO2 mixtures reduces the NOx emission rate from fuel-N, which indicates the potential chemical effect that CO2 may have in NOx formation/reduction chemistry. In particular, the CO2 effect on the heterogeneous reactions that may lead to NO reduction has not yet been investigated. The effect of high CO2 and O2 levels on heterogeneous formation and destruction of NO is the focus of the fundamental studies.

For the experimental work, a single-particle, fluidized bed reactor (SFBR) has been constructed for use in studying char oxidation rates and fuel nitrogen conversion as a function of temperature, PO2 and PCO2. Studies are being carried out with model chars, as well as coal and coal-derived chars. Model chars for use in the char-N conversion studies are created from polyacrylonitrile, with subsequent characterization of the N functionalities in the char by XPS.

The computational work is focused on the interactions of high levels of CO2 found in oxycoal combustion, with fuel nitrogen present in the coal. The calculations are being performed using the Gaussian 03 package, using several different N functionalities in the char, as well as various representative gaseous environments.

Pilot-Scale Studies

Pilot-scale (300 kW) oxycoal CFB data is being obtained for use in validation of the Multiphase Flow with Interphase eXchanges (MFIX) computer model being developed by U.S. DOE/NETL.

Measurements include: pressure, temperature and bed voidage mapping, and as well as gas composition and velocity mapping.

This presentation will provide an overview of the fundamental and pilot-scale studies, along with progress and results achieved to date.

REFERENCES

1. Park J., et al., Energy & Fuels, 2007. 21: p. 121-129.

2. Croiset E. and Thambimuthu K.V., Fuel, 2001. 80: p. 2117-2121.

3. Hu Y., et al., Fuel, 2000. 79: p. 1925-1932.

676a Modeling the Carbon Dioxide Sequestration Process: Technology Gaps and Validation

Milind Deo, **Prashanth N. Mandalaparty**, and Zhiqiang Gu. Chemical Engineering, University of Utah, 72 S Central Campus Drive Merill Engineering Building Rm 3290, Salt Lake city, UT 84112

A great deal of effort has been expended in the last decade to model carbon dioxide (CO2) sequestration processes. Data on a variety of scales and understanding of a number of processes is necessary in order to develop a consistent understanding of the important mechanisms of CO2 transport and reactivity with the aim of understanding the long-term fate of CO2 in underground formations. The reactions are often described using equilibrium models, which may or may not be relevant on the time scale of the simulation. In the first part of this paper, we evaluate the various equilibrium and kinetic models used and their implication on the ultimate fate of CO2. It has often been observed in natural CO2 reservoirs that the type of mineralogical changes observed in the laboratory or predicted in models are only partially seen in the field. The reasons for this and possible model modifications necessary to adjust for these phenomena are discussed next. Impact of the mineralogical changes on injectivity are discussed next. The transport and of CO2 in the reservoir depends on reservoir properties, on rock-fluid interactions (capillary pressures and relative permeabilities), and on the interactions of CO2 with the fluid (brine) and the rock. We take a comprehensive look at all these effects together in light of our findings on reactivity and injectivity. These findings are examined in the context of the validation data available on CO2 injections on a variry of scales.

676b Effects of Permeability Heterogeneity on CO2 Sequestration In Brine Formation

Si-Yong Lee, Weon Shik Han, Chuan Lu, Brian McPherson, and Richard Esser. Energy & Geoscience Institute / Department of Civil and Environmental Engineering, University of Utah, 423 Wakara Way Suite 300, Salt Lake City, UT 84108

We present results of a study focused on the impacts of geologic heterogeneity on CO2 sequestration in deep saline formations. For systematic spatial variations in two-dimensional permeability (k) fields, we implemented 13 scenarios, including homogeneous, random, homogenous with a low-k inclusion, and anisotropically correlated random fields. Unlike previous studies solely based on a single realization, equally probable 10 realizations to each scenario except for the homogeneous cases were generated with sequential Gaussian simulation method and then served to the input for the numerical model of CO2 migration for a 200-year simulation period. To characterize the simulated CO2 plume behavior and associated trapping mechanisms, we divided and quantified the trapped CO2 as mobile-, residual-, and aqueous-trapped CO2. In addition, the first and second moments of CO2 plume distribution were examined to explore the both lateral and vertical CO2 displacement behavior according to the different geologic complexity. Simulation results from both homogeneous and uncorrelated random fields show distinct positive correlation between the effective vertical k and residual CO2 trapping. More interestingly, our results with the anisotropy in correlation length structure indicate that the degree of horizontal connectivity in log-normally distributed k fields tends to correlate with the amount of residual trapping only when the k anisotropy ratio of horizontal to vertical k was increased by one order of magnitude from the isotropic permeability. In contrast, no such strong relationship between the horizontal correlation length and residual trapping was found for isotropic permeability. This suggests that a brine reservoir with a longer lateral connectivity and greater k anisotropy ratio can better serve as a CO2 sequestration site in terms of residual trapping and buoyancy-driven migration.

676c Including Long-Range Electrostatic Interactions In Carbon Dioxide Sequestration Models

Shela J. Aboud, Yangyang Liu, and Jennifer L. Wilcox. Department of Energy Resources Engineering, Stanford University, 367 Panama St., Green Earth Sciences 077A, Stanford, CA 94305-2220

The capture and sequestration of carbon dioxide in deep coal beds including enhanced methane recovery is an attractive option for reducing atmospheric emissions. Microscopic simulation approaches such as Monte Carlo and Molecular Dynamics are powerful techniques that can be used to probe the physical and chemical interactions between the carbon dioxide, methane and coal-pore surfaces. Information, including adsorption isotherms, diffusivity, and permeability can be extracted using these computational methods and then can be used to calibrate larger-scale macroscopic models. These modeling techniques have been widely used on idealized pore structures, generally represented by smooth graphite surfaces or carbon nanotubes. A key issue preventing the implementation of these simulation techniques to study the heterogeneous structures which are more representative of the actual coal surfaces is the lack of efficient long-range force-field schemes.

The real surface of the coal pores includes both polar and nonpolar functional groups making the interactions between the walls and the molecules highly anisotropic and limiting the accuracy of the commonly used 10-3-4 Steele potential[1]. In this work an efficient force field scheme will be introduced to account for the anisotropic distribution of the electrostatic forces in real coal pores. Although novel to this field, the scheme is based on the particle-particle particle-mesh (P3M) technique and was originally developed and validated for use in simulating ionic transport in biological ion channels[2]. Within the P3M approach, the forces are divided into a short-range force that accounts for neighboring particle interactions and a smoothly-varying long-range force.

Short-Range Interactions: The short-range, particle-particle force is defined within a small sphere surrounding each particle and includes a Coulomb term, a van der Waals term and a reference term that corrects for any double counting due to the overlap of the short- and long-range domains. The van der Waals interactions are typically included using Lennard-Jones potentials. In this work, to account for the possibility of stronger interactions, i.e., chemisorption, data generated from first-principle calculations using Gaussian03 or VASP will be incorporated. These data contain the interaction information, such as potential energy and partial charges, as a function of distance between the surface groups on the coal and CO_2 and CH_4 molecules. To reduce computational time interactions can be stored in look-up table and accessed during run-time.

Long-Range Interactions: The second part of the P3M method, the particle-mesh term, contains the long-range interaction which is from the Coulomb force between far charges and is computed on a finite difference mesh in real space using Poisson's Equation and the iterative multigrid method. The mulitgrid method is one of fastest and efficient iterative solvers available for solving large sparse matrix equations, scaling as O(N)[3]. The approach can be shown in the following steps:

1. Assign particle charges to mesh points in the computational domain

2. Using the approximated charge density, calculate the electrostatic potential at each mesh point by solving Poisson's equation using the multigrid method

3. Calculate the electric field at each mesh point by differentiating the potential numerically

4. Interpolate the force from the mesh point to particle

It is important to note that this approach does not assume periodic boundary conditions, making it more flexible for simulating anisotropic systems.

The force-field scheme described will be validated against the Steele potential for ideal smooth graphite layers and pores structures and the error associated with excluding the long-range portion of the electrostatic interactions will be examined within more realistic pore structures that include polar functional groups representative in the coal matrix. Extensive timing comparisons will be made to illustrate the effectiveness of the approach for use in Monte Carlo and Molecular Dynamics simulations.

[1] W. A. Steele, Sur. Science, 36 (1973) 317

[2] D. Marriero, M. Saraniti, S. Aboud, "Brownian Dynamics simulation of charge transport in ion channels",, J. Phys.: Condens Matter 19 (2007) 215203.

[3] W. Hackbush, Multi-Grid Methods and Applications, Springer-Verlag, Berlin (1985).

676d CO2 Sequestration: Temperature and Gas Compositional Effects on the Kinetics and Mineralogical Reactions

Prashanth N. Mandalaparty[1], Kyeongseok Oh, Milind Deo, and Joseph Moore[2]. (1) Chemical Engineering, University of Utah, 50 S Central Campus Drive Merill Engineering Building Rm 3290, Salt Lake city, UT 84112, (2) Energy and GeoSceince Institue, University of Utah, 423 Wakara Way suite 300,, Salt Lake City, UT 84108

The alarming rate of the CO2 emissions, which is eventually the main cause for global warming, is increasing the concern for the growing need to look for different means of depositing the atmospheric carbon dioxide. Geological sequestration of CO2 seems to provide an answer to this problem, where CO2 is concentrated and pumped into deep saline aquifers, coal formations or depleted oil reservoirs. Despite the care taken during the purification and isolation of flue gases from coal fired power plants, there are traces of other gases such as SO2 remain in the flue gas stream. The reactive behavior of pure CO2, CO2+ SO2 within this geologically and chemically complex environment is being studied. The experimental apparatus consists of a series of 6 isolated high pressure reactors operated at different conditions and with different feed gas compositions to study and observe the mineralogical changes in both the rock and water chemistries. A 3g rock sample was prepared by mixing equal amounts of finely ground (100μ m) quartz, calcite, andesine, dolomite, chlorite and magnesite, which is usually called arkose or dirty sand. Brine was prepared from laboratory grade sodium chloride with a concentration of 2g/30cc. The long term batch experiments with pure CO2 as feed gas were carried out at 100C and 200C as different sets in four separate reactors to evaluate the dependence of kinetics and the mineralogical changes on temperature. The operating pressure was constant at ~2000PSI. The system operating at 200C and 2000PSI was held at these conditions for a period of 37 days before depressurizing. The system held at 100C and 2000PSI was depressurized after a period of 62 days. The mineralogical changes in the rock were analyzed using SEM/EDS and XRD analysis. The brine chemistry was analyzed using ICP-MS for cations and ion chromatography for anions. Each mineral in this complex geological environment shows evidence of participating in the geo-chemical reactions. On analyzing the system which was at 200C and 2000PSI for changes in the rock compositions, interesting results were found. Layers of calcite were seen growing on the surface of the arkose. Analcime deposits are almost omnipresent either occurring as large connected aggregates or as deposits on the surfaces of other minerals. Ankerite deposition was observed as an amorphous mass intergrown with the starting minerals. The growth of these new phases indicates the complex geo-chemical interactions taking place. The temperature dependence of these reactions is being evaluated by analyzing the set of experiments carried out at 100C and 2000PSI. The changes in the brine chemistry for both sets are being analyzed. The rates of these reactions will be computed by computing the changes in the compositions of the initial and the final rock samples provided by the XRD analyses and its dependence on temperature will be evaluated. The experiments with 10%SO2 and 90% CO2 are being carried out in which anhydrite precipitation and dolomite deposition are expected due to increased acidity and sulphate concentration. The data has been analyzed in light of the published equilibrium and kinetic models. These experiments will help in evaluating the gas compositional effects on the kinetics and the mineralogical reactions.

676e Numerical Investigation on Low-Viscosity-Ratio Drop Breakup In a Contraction Flow

Guangdong Zhu and Andrea A. Mammoli. Mechanical Engineering Department, University of New Mexico, MSC01-1150, Albuquerque, NM 87131

The breakup of low-viscosity-ratio fluids can often be encountered in the process of CO2 storage through saline aquifer but remains a challenging problem due to its complexity in physics. By using an accurate indirect boundary element method, the breakup of a low-viscosity-ratio isolated drop in a contraction flow at a vanishing Reynolds number is investigated numerically. A practical mathematical method is constructed to detect the asymptotic behavior of maximum curvature at pinching point and used to predict the occurrence of a pinch-off. The 3-D numerical simulation presented here can accurately capture not only the primary breakup of a low-viscosity-ratio drop, but also the secondary breakup and presence of a set of satellite drops. The results agree qualitatively with laboratory experiments but provide more details to understand the process of low-viscosity-ratio drop breakup in a confined outer flow.

676f Uncertainty Modeling In Sequestration Applications

Milind D. Deo, Zhiqiang Gu, and Prashanth N. Mandalaparty. Chemical Engineering, University of Utah, 50 S. Central Campus Dr., MEB, Rm. 3290, Salt Lake City, UT 84112

Quantitative assessment of geologic uncertainty has been of great interest in the last few years, and that interest is growing with the advent of new carbon sequestration testing programs. In the oil industry, several different approaches have been used to obtain probability distribution functions (PDFs) of desired parameters, such as hydrocarbons in place, recovery factors, etc. In CO2 sequestration one might want to determine critical fault properties that could lead to hydrostratigraphic trapping failure or to thickness variations of the seal that could lead to seal breach. The two most prominent methods for developing effective PDFs are:

• Experimental design based methods

• Bayesian probabilistic formalism

In general, experimental design methods are based on generating simpler response surfaces using selected rigorous simulations (e.g., of reservoir models with appropriate fully coupled phenomenological process models) followed by Monte Carlo simulations to identify the parameters that are most critical in affecting the targeted outcome. Sufficient data to validate models of trapping mechanisms and associated failure modes will not necessarily be available from any previous as well as ongoing field tests. For example, very few new wells that could provide calibration data, including log and seismic data, will be drilled in most ongoing or planned field demonstrations. In these situations where information is sparse, the Bayesian framework can be used to obtain probability distributions of critical parameters. In this approach several conceptual geologic models are constructed based on all available data. Each scenario is then depicted using a spatial variability model. It is recognized at this stage that several choices of spatial variability models (variograms, for example) are available. The global estimate of a given target variable is constructed using the interpretation of the quantitative data under a given geologic scenario. Once the uncertainty in a given parameter is quantified, experimental design tools can be used to design monitoring programs with an aim of reducing uncertainty. A general framework for determining and quantifying uncertainty in carbon dioxide sequestration applications is built and some examples are shown.

Chemical-Looping Combustion of Coal with Metal Oxide Oxygen Carriers

Ranjani Siriwardane*[1], Hanjing Tian[1,2], Thomas Simonyi[1,2], and George Richards[1]

[1]*U.S. Department of Energy, National Energy Technology Laboratory, 3610 Collins Ferry Road, P.O. Box 880, Morgantown, WV 26507-0880*

[2]*Parsons, PO Box 618, Pittsburgh, PA 15129*

* Corresponding Author
Ranjani Siriwardane
E-mail: ranjani.siriwardane @netl.doe.gov

INTRODUCTION

Chemical looping combustion (CLC) has been suggested as an energetically efficient method for producing high-purity CO_2 from combustion of fuel gas. This is an entirely new combustion technology, which involves the use of a metal oxide as an oxygen carrier that transports oxygen from the air to the fuel, avoiding direct contact between fuel and air.

The CLC system is composed of two reactors, an air and a fuel reactor. In the fuel reactor, the fuel in the gaseous form reacts with the metal oxide.

$$\text{Fuel} + \text{Metal oxide} \rightarrow CO_2 + H_2O + \text{Metal or reduced form of metal oxide} \quad (1)$$

In the air reactor, the metal or a reduced form of metal oxide is oxidized to form metal oxide.

$$\text{Metal or reduced form of metal oxide} + O_2 \rightarrow \text{Metal oxide} \quad (2)$$

The regenerated metal oxide is ready to initiate a second cycle. The exit gas from the fuel reactor contains CO_2 and H_2O and after condensation of the water produces a pure CO_2, which can be utilized for sequestration. The significant advantage compared to normal combustion is that concentrated CO_2 not diluted with N_2 is obtained in a pure state without expending any major energy required for separation. This could enable combustion devices that produce thermal energy with CO_2 inherently ready for geological sequestration.

It is important to note that the heat of reaction of both reactions (1) and (2) must be equal to the normal combustion heat of reaction between the fuel and oxygen. However, depending on the choice of metal/metal oxide, it is possible that one of the reactions can be endothermic. In a practical application, this means that heat must be transferred from the exothermic reactor to sustain the endothermic process. This can be accomplished, for example, by using the heat capacity of the circulated metal to carry heat between the reactors, and will be mentioned later.

Only a few studies[1-3] have been conducted on the combustion of solid fuels such as coal. There are many challenges associated with direct CLC of coal using oxygen carriers. The oxygen carrier should have sufficient rate suitable for various reactor systems, sufficient oxygen release capacity facilitating the coal-oxygen carrier interactions, stable reactivity during multiple cycles, high attrition resistance, and low reactivity with ash and other contaminants.

In the present work, various metal oxides—CuO, Fe_2O_3, NiO, Mn_2O_3 and Co_3O_4— were tested as oxygen carriers for direct combustion of coal in the presence of either nitrogen or CO_2. Thermo gravimetric analysis (TGA) was used to determine the rates of reaction and percentage of combustion/oxidation. The calculations of the amount of solid that will be required for designing a large-scale unit will also be discussed.

EXPERIMENTAL

TGA experiments were conducted in a Thermo gravimetric Analyzer (TA Model 2050). Illinois #6 coal was used in this study with particle size of 100 microns. The composition of coal (proximate analysis data) is listed in Table 1.

Table 1: Composition of coal

	Concentration (%)
moisture	7.44
Volatiles	37.97
ash	12.45
H	4.63
C	67.32
N	1.3
S	4.81
O	9.49

The coal was mixed with metal oxides (dry mixing) and the weight ratio was about 0.2 grams of coal to 3-4.5 grams of metal oxide, which corresponds to a volume ratio of metal oxide to coal of about 2.6-3.5 to 1. Thus, excess metal oxide was used in these early studies. Future work will define minimum stoichiometries needed for complete combustion. About 200 mg of the coal/metal oxide mixture was heated in a quartz bowl from ambient to 900-1000 °C at a heating rate of 15 °C/min in either pure nitrogen or pure CO_2 at a flow rate of 40 sccm. The sample then was kept isothermal at either 900 or 1000 °C for about 80 minutes; afterwards, air was introduced at 40 sccm for about 40 minutes.

The fractional conversions (fractional reduction and fractional oxidation) were calculated utilizing the TGA data. The fractional conversion (X) is defined as

$$\text{Fractional Reduction } (X_r) = (M_o - M)/(M_o - M_f)$$

$$\text{Fractional Oxidation } (X_o) = (M - M_f)/(M_{oxd} - M_f)$$

M = *Instantaneous weight of metal oxide-coal mixture at a given time*

M_o= *Initial weight of metal oxide-coal mixture prior to the reaction*

M_f= *Final weight of metal oxide-coal mixture after reaction in either* N_2 *or* CO_2

(i.e., reduced metal + ash + unreacted coal)

M_{oxd} = *Weight of completely oxidized sample after introducing air*

The fractional conversion data as a function of time was fitted to obtain the polynomial regression equation. The global rates of reactions (dX/dt) at different fractional conversions (X) were calculated by differentiating the polynomial equation.

Percentage of combustion and percentage of oxygen uptake were obtained utilizing the weight change data from the TGA using the following equations:

Percent combustion = (actual weight loss of coal from TGA/Theoretical weight of coal in the sample) × 100

Percent oxygen consumption = (experimental oxygen consumption / theoretical capacity of oxygen presenting the metal oxide) × 100

Notice that the percent combustion is defined with respect to the sample test time (80 minutes at 900 or 1000 ^{0}C).

RESULTS AND DISCUSSION

The reaction mechanisms of the coal-oxygen carriers are not well defined. It is reported that the gaseous products from coal pyrolysis may initiate the reaction with oxygen carrier[2]. The major goal of the CLC process is to obtain a concentrated CO_2 stream after the combustion process; therefore, the reaction in the fuel reactor has to be conducted in the presence of CO_2 or possibly steam (followed by water condensation). It is hypothesized that the CO produced by coal gasification by CO_2 may initiate the reaction with the oxygen carrier in the fuel reactor. It has also been reported that the gaseous oxygen released from the oxygen carrier could initiate the reaction at high temperatures.[3] Additional thermodynamic studies have reported that direct metal oxide-carbon interaction is also possible with some metal oxides.[2]

a. Thermo Gravimetric Analysis - Base Line Tests with Coal without the Metal Oxide

Baseline tests were conducted with pure coal utilizing N_2 and CO_2. The TGA data when coal was heated in N_2 up to 1000 °C are shown in Figure 1. A small peak starting at 82 °C corresponds to dehydration of coal. A larger peak starting at 250 °C with a maximum at 435 °C is also observed. This indicates that the coal de-volatilization was initiated at 250 °C with the maximum rate at 435 °C. The weight loss due to coal de-volatilization in N_2 was about 45% and no additional weight losses were observed during the heating process in N_2. This weight loss matches the specified volatile matter from the coal proximate analysis as shown in Table 1.

Figure 1: TGA Profile of Coal in N_2 without Metal oxide

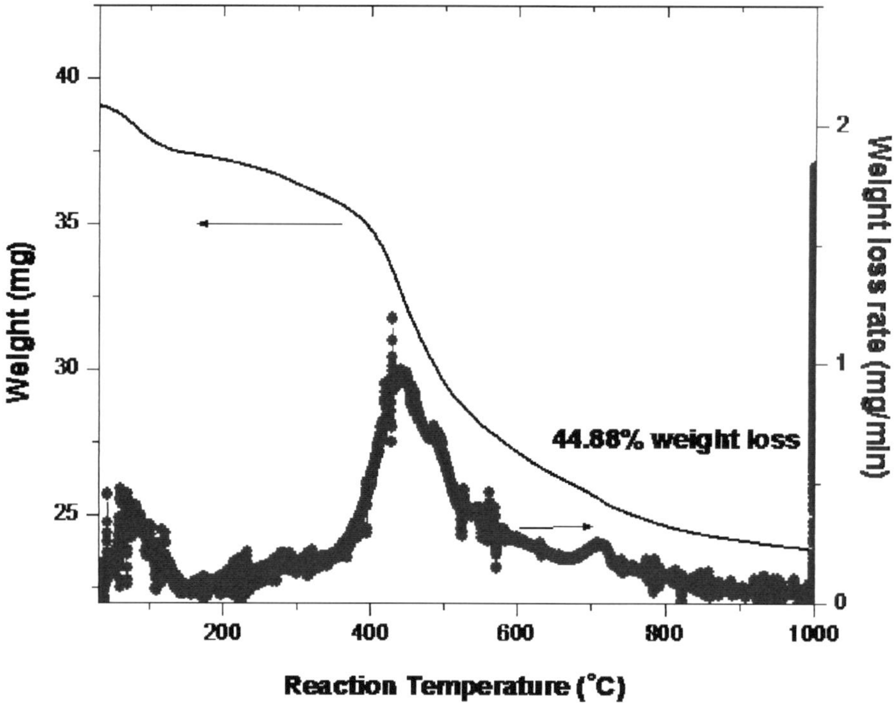

Additional weight loss of 55% due to combustion reaction was observed when air was introduced at 1000 °C subsequent to the N_2 introduction. When coal was heated in CO_2 the weight loss profile and rate data were similar to that with N_2. This indicated that the presence of CO_2 did not have any significant effect on the process.

b. Thermo Gravimetric Analysis – Tests with Coal and Metal Oxides in N_2 and CO_2

The weight loss profile and corresponding rate data of coal-CuO mixture during the heating in N_2 up to 900 °C is shown in Figure 2. There is a continuous weight loss with the coal-CuO sample during the heating process which indicated that the CuO contributed to the combustion of coal after the initial pyrolysis reaction. This weight loss corresponds to 100% coal combustion with CuO in the presence of N_2. The rate data shown in Figure

2 indicated the presence of peaks at 425 °C and 708 °C. The major peak at 708 °C is due to coal combustion by oxygen from CuO and the coal combustion appears to initiate around 500 °C. The outlet CO_2 concentration from the TGA during reaction measured utilizing a mass spectrometer is also shown in Figure 2.

Figure 2: TGA and Mass Spectral Data for Coal Combustion with CuO

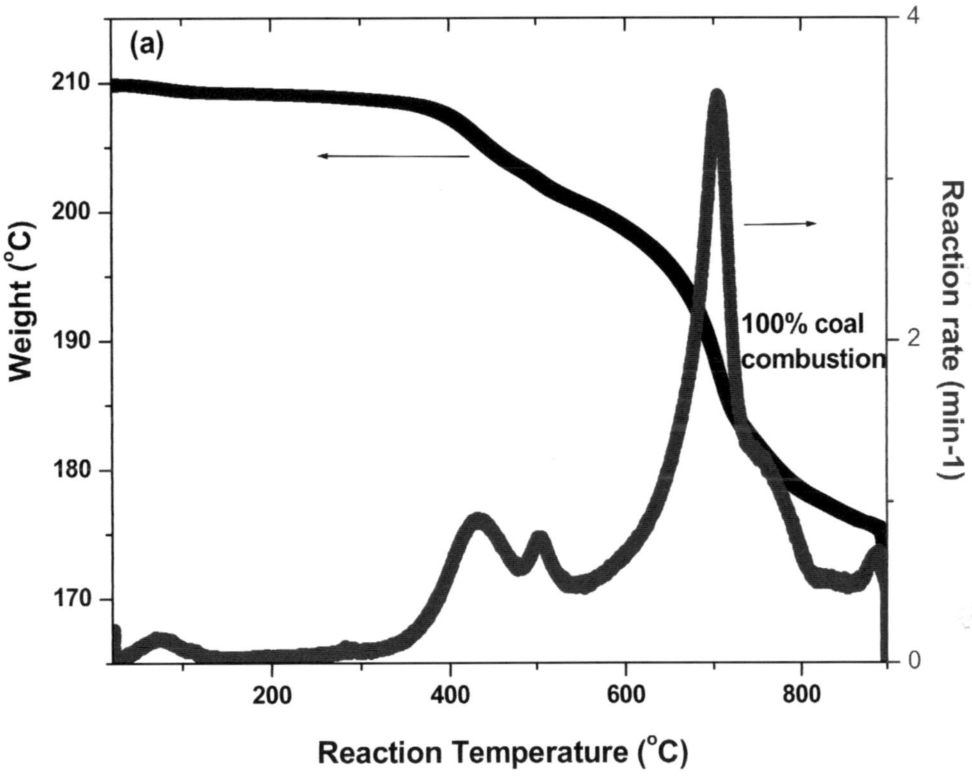

The rate data from the TGA measurements and the outlet CO_2 profile appear to be very consistent. The TGA data when air was introduced after the reduction reaction with coal-CuO is shown in Figure 3. A weight gain due to oxidation was observed and reduced copper was fully oxidized. The rate of oxidation is significantly higher than that for the reduction reaction.

Figure 3: Oxidation Segment of CuO in Air at 900 °C

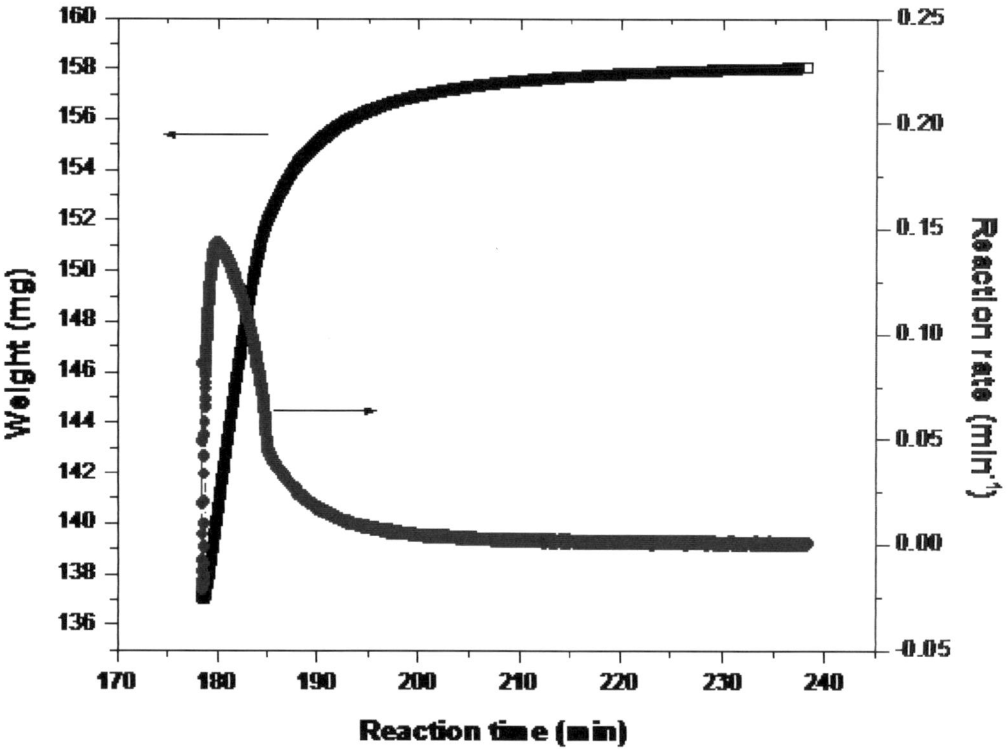

The goal of CLC reaction of coal with metal oxides is to obtain a pure stream of CO_2 after combustion; consequently the combustion reaction of coal with metal oxide should be performed in the presence of CO_2. TGA experiments were also performed with coal and CuO, NiO, Fe_2O_3, Mn_2O_3, and Co_3O_4 in the presence of both N_2 and CO_2; these data are shown in Table 2.

Table 2: Coal-CLC Data of Various Metal Oxides in N_2 and CO_2. Combustion temperature indicates the temperature where the highest combustion rate was observed. Percent combustion is defined with respect to the sample test time (80 minutes at 900 or 1000C)

		Combustion				Oxidation		
sample	gas media	Comb. Temp. (°C)	Comb. Rate (min^{-1})	% combustion	ΔH (kJ/mol)	Oxi. Rate (min^{-1})	oxygen uptake (%)	ΔH (kJ/mo
	CO_2	703	0.098	100		0.172	98.6	
CuO	N_2	708	0.083	100	-96.5	0.175	99.2	-156
	CO_2	993	0.061	73.05		0.84	77.5	
NiO	N_2	1000	0.017	68.4	75.2	0.82	71.6	-327.7
	CO_2	973	0.055	94.9		0.77	93.7 Fe(II)	
Fe_2O_3	N_2	977	0.05	91.6	79.2	0.78	90.6 Fe(II)	-347.4
	CO_2	905	0.011	76.76		0.42	72.2Mn(II)	
Mn_2O_3	N_2	978	0.01	71	-36.1	0.38	68.3 Mn(II)	-216.4
	CO_2	781	0.096	83.3		1.74	78.2 Co(II)	
Co_3O_4	N_2	781	0.096	83	-8.6	1.74	78.0 Co(II)	-243.9

It is interesting to note that the lowest combustion reaction temperature at 700 °C and the highest combustion rate were observed for a CuO-coal system. In addition, the combustion reaction with CuO is exothermic (ΔH=-96.5 KJ/mol) which indicates that the heat transfer from the oxidizer to the reducer is not necessary for the reaction to proceed. This is an advantage of the CuO-coal system. The percentage of combustion and oxidation were close to 100% for the CuO which indicated that the complete coal combustion reaction could be obtained with the CuO and reduced copper could be completely oxidized at 900 °C. In the presence of CO_2, the rate of combustion for coal-CuO was slightly higher than that with N_2, but there was no significant effect due to CO_2 for a CuO-coal system.

For Fe_2O_3-coal system, (maximum) combustion rate takes place at a higher temperature (973-977 °C) than that with CuO. It is possible to achieve about 95% combustion with Fe_2O_3. The percentage of oxidation was about 77 % when the oxidation state change of Fe was assumed to be from Fe(III) to Fe(II) during the CLC reaction. The presence of CO_2 did not have a significant effect on both the reaction rate and the combustion temperature of the Fe_2O_3–coal system. The heat of the reduction reaction for an Fe_2O_3-coal system is endothermic (ΔH=79.2 KJ/mol) and heat transfer from the oxidizer to the reducer will be necessary for a CLC process with Fe_2O_3, which could be achieved utilizing a heat-carrying inert material, or using the heat capacity of the circulating iron. The combustion rate for Fe_2O_3 appears to be lower than that with the CuO system, but the oxidation rate appears to be higher with Fe_2O_3.

The characteristics of a NiO-coal system were similar to the Fe_2O_3-coal system as shown in Table 1. For NiO-coal the combustion reaction—which is endothermic (ΔH=75.2 KJ/mol)—also takes place at higher temperature of 993 °C with a lower reaction rate than that with CuO-coal system. However, for NiO the presence of CO_2 had a significant positive effect on the combustion properties. For Mn_2O_3-coal, the combustion reaction took place at 900 °C but showed the lowest combustion rate as compared to that with the other four metal oxides, but the reaction is slightly exothermic (ΔH=-36.1 KJ/mol). The percentage of combustion was similar to that with NiO. For Mn_2O_3, the reduction-oxidation during the CLC reaction appears to occur between the oxidation states Mn(III) and Mn(II).

The Co_3O_4-coal system appears to be similar to that of CuO with a low combustion temperature of 781 °C, but the oxidation rate was the highest for the Co_2O_3. The heat of reaction is slightly exothermic (ΔH=-8.6 KJ/mol), yet the percentage of combustion was lower than that with CuO-coal.

From the five metal oxides tested, CuO appeared to have the best combustion characteristics. It had the lowest combustion temperature, highest combustion rate, and highest percentage combustion. The CuO and Cu metals have a lower melting point (~1000 °C) than other metal oxides. However, the coal combustion reaction with CuO takes place at a lower temperature (700 °C) than the melting point of copper oxide and may be a potential candidate for the coal CLC process.

An obvious question is determining how solid metal oxide reacts with the solid coal? It may be possible that coal volatiles initiate the reaction between the metal oxide and the solid fuel, with the remaining char gasified by combustion products. However, the coal de-volatilization reaction takes place around 450 °C and the concentrations of volatile combustion products are very low at the coal combustion temperatures with the metal oxides. As a result, it is not clear whether the volatile combustion products are required to initiate the reaction, and more work is needed to understand the exact mechanism responsible for complete coal combustion.

There are numerous practical issues that must be addressed to develop a complete chemical looping combustion system. Ash separation from the reactor might be accomplished by the density difference between the ash and metal. The quantity, cost, and durability of the metal oxide are all important considerations. The data presented here are being used to determine which approaches may be successful in a practical device aimed at solid fuel combustion. A key concern is the circulation of hot solids between the reactors. It is helpful to note that existing fluidized catalytic cracking (FCC) reactors have demonstrated significant commercial experience with solid catalyst circulation rates typically 35 tons per minute (or higher), operating with regenerator temperatures of 700C[4]. While the chemical looping process described here is different, the commercial experience of handling large flows of hot solids provides some basis for process development.

CONCLUSIONS

Metal oxides CuO, Fe_2O_3, NiO, Mn_2O_3, and Co_3O_4 showed the ability to provide oxygen for direct combustion of coal. CuO showed the lowest temperature, the highest percentage (100%), and the highest rate for the coal combustion. Reduced Cu metal during the coal combustion can be fully re-oxidized to the copper oxide. The highest oxidation rate was observed with Co_3O_4. Coal combustion reactions with Fe_2O_3, NiO, and Mn_2O_3 took place at higher temperatures (>900 $^\circ$C) than that with CuO and Co_3O_4. The presence of CO_2 had a significant positive effect on the coal combustion reaction with NiO. A process of direct coal combustion may be feasible with metal oxides, but it will be necessary to develop an efficient solid circulation process and an ash/metal-oxide separation process. Ash generated during the reaction has a lower density than that of metal oxides and can be separated utilizing density separation.

REFERENCES

1. Andrus, H.E. Jr., and McCartney, M.S., "Hot Solids Gasifier with CO_2 Removal and Hydrogen Production", U.S. Patent 2,083,658 B2, Aug. 2006.

2. Cao Y., and Pan W., Energy and Fuels 2006, 20, 1836-1844.

3. Berguerand, N., and Lyngfelt, A., *International Journal of Green House Gas Control* 2, 2008, 169-179.

4. Equipment Design and Cost Estimates for Small Modular Biomass Systems, Synthesis Gas Cleanup, and Oxygen Separation Equipment, NREL/SR-510-39943, available www.osti.gov/bridge, DOE contract DE-AC36-99-GO10337.

686b Ni/la-Al2O3 Oxygen Carrier for Fluidized Bed Chemical-Looping Combustion

Mohammad M. Hossain, David Lopez, Jose Herrera, and Hugo de Lasa. Chemical and Biochemical Engineering, The University of Western Ontario, London, ON N6A 5B9, Canada

Chemical-looping combustion (CLC) is a novel concept in the area of CO2 capture. In this process, fossil fuel combustion is carried out in two fluidized bed reactors: a fuel reactor and a regenerator. In the fuel reactor, the fuel is burnt using a solid oxygen carrier. Thus, the exhaust from this reactor contains only CO2 and H2O. Pure CO2 can be obtained when the H2O vapor is condensed. The reduced carrier is reoxidized in regenerator with air and cycle back to fuel reactor. CLC also helps minimizing NOx emissions, since the fuel is burned in the absence of nitrogen without a flame. This study deals with the development of a La modified Ni/La-ãAl2O3 oxygen carrier suitable for a CLC process.

In CLC, the most important characteristic of an oxygen carrier is its reactivity and stability under cyclic operation. To investigate these matters successive TPR and TPO experiments were developed in the context of the present study. In cyclic TPR/TPO experiments, the reduction and oxidation profiles of Ni/La-ãAl2O3 sample remains unchanged and the amount of Ni reduction also shows a stable behavior. Approximately 93 % nickel conversion was achieved using the La modified sample, while the value was 76 % range for the unmodified carrier. Therefore, the addition of La aids the formation of easily reducible nickel oxides minimizing nickel support interaction and formation of non-reactive nickel aluminates.

The pulse chemisorption results further confirmed the stable behavior of the sample in consecutive redox cycles. Although both unmodified and modified samples showed a stable metal dispersion with successive redox cycles, a higher dispersion was observed in the modified sample versus the unmodified samples, from 2.5 % to 0.9%. This indicates that La helps to increase the dispersion of nickel. Thus, the addition of La alters the metal surface modifying the degree of interaction between Ni and alumina support, maintaining a consistent metal dispersion during the repeated redox process. It is also apparent that the metal crystal size of the Ni/La-ãAl2O3 sample remained unchanged over repeated cycles, an indication of the absence of agglomeration of nickel crystals.

686c Effect of Sulfur Contaminants on the Performance of Oxygen Carriers In Chemical Looping Combustion

Rahul D. Solunke and Götz Veser. Chemical Engineering Department, 1249 Benedum Hall, University of Pittsburgh, 3700 O'hara Street, Pittsburgh, PA 15261

Chemical looping combustion (CLC) is an emerging technology for clean energy-production from fossil and renewable fuels. In CLC, an oxygen carrier (typically a metal) is first oxidized with air. The hot metal oxide is then reduced in contact with a fuel in a second reactor, thus combusting the fuel. Finally, the reduced metal is transferred back to the oxidizer, closing the materials "loop". In this way, CLC produces sequestration-ready CO2-streams without expensive air separation and hence without significant energy penalty. Combined with sequestration, CLC thus allows high-efficiency, CO2 emissions-free combustion of fossil fuels, or combustion processes with negative CO2-footprint from biomass-derived fuels. However, CLC suffers from insufficient stability of typical metal-based carrier materials at high-temperature redox conditions.

We previously demonstrated that the embedding of metal nanoparticles into a ceramic matrix can result in unusually active and sinter-resistant nanocomposite materials which combine the high reactivity of metals with the high-temperature stability of ceramics. The resulting materials show very fast oxidation and reduction kinetics, and were stable in multiple redox cycles in the temperature range of interest for chemical looping (~700 – 1000oC).

Here, we report on the effect of sulfur contamination in the fuel stream on the performance of different oxygen carriers in CLC. Most of the published work on oxygen carriers to-date has focused on their performance in "idealized", contaminant-free fuel streams. However, realistic fuel streams contain significant amounts of contaminants, in particular sulfur (mainly in the form of H2S), which can significantly impact on the performance of CLC via sulfidation of the carrier material. Furthermore, metal sulfides often have lower melting points than the corresponding metals or metal oxides, and thus put an additional constraint on the operating temperature of a CLC process.

We studied the effect of H2S contamination on the performance of a nanostructured Ni-barium hexaaluminate carrier at different temperatures in CLC. Thermodynamic calculations were combined with experimental investigations. We found that in addition to sulfidation of the metal carrier, sulfidation of the support structure also needs to be considered. While metal sulfidation is a reversible process, the sulfidation of the support is irreversible and can result in structural changes in the carrier material. We are currently further investigating the effect of sulfur contaminants on the kinetic behavior of these carriers in cyclic-TGA (thermogravimetric analysis) experiments with coal-derived synthesis gas.

Preparation and characterization of the carriers before and after exposure to sulfur contaminants, as well as their performance in CLC will be discussed in detail in the presentation.

A DUAL CIRCULATING FLUIDIZED BED (DCFB) SYSTEM FOR CHEMICAL LOOPING PROCESSES

Tobias Pröll, Philipp Kolbitsch, Johannes Bolhàr-Nordenkampf and Hermann Hofbauer
Vienna University of Technology, Austria

Abstract

A fluidized bed system combining two circulating fluidized bed reactors is proposed and investigated for chemical looping combustion using metal oxides as oxygen carriers. The global solids loop starts in one of the two reactors (primary reactor) where solids are entrained, separated from the gas in a cyclone and sent to the other reactor (secondary reactor) via a fluidized loop seal. From the secondary reactor, the solids are flowing back into the primary reactor via a second loop seal connecting the bottom regions of the two reactors. The secondary reactor features a circulation loop in itself (secondary reactor cyclone and loop seal) and may be optimized with respect to good gas-solid contact and low particle attrition. The global circulation rate can be effectively controlled by staged fluidization of the primary reactor. The direct hydraulic communication of the two circulating fluidized bed reactors via the lower loop seal allows stable solids distribution in the system. A 120 kW fuel power laboratory unit has been designed for chemical looping combustion of natural gas. The air reactor, where oxygen carrier particles are oxidized, is designed as the primary reactor. The fuel reactor, where the oxygen carrier serves to oxidize gaseous fuel, is designed as the secondary reactor. The reactor dimensions have been determined considering fast fluidization in the air reactor and turbulent fluidization in the fuel reactor. Experimental results with a NiO-based oxygen carrier show that fuel conversion increases with operating temperature. For the investigated system, the CO_2 yield from CH_4 is 0.94 at 950 °C (1223 K). It is expected that fuel conversion can be further improved by increasing the height of the riser reactors.

Introduction

Carbon capture and storage is the key mid-term strategy to limit the carbon dioxide concentration in the atmosphere [1]. Different approaches are currently discussed in order to capture CO_2 from combustion processes. Most of these approaches require gas separation steps: either CO_2 separation (pre- and post combustion capture) or O_2 separation from air (oxyfuel combustion). Chemical looping combustion (CLC), however, is a way of indirect combustion where fuel and air are never mixed [2]. Metal oxides are used to selectively transport oxygen from air to fuel in the solid phase. If a suitable metal oxide is used as oxygen carrier, the CLC system can be operated in a way that the exhaust gas of the fuel reactor ideally consists of CO_2 and H_2O only and allows for subsequent water condensation, compression, and storage of CO_2. Therefore, CLC is discussed as one of the very energy-efficient approaches towards carbon capture from power production or fuel upgrading.

Chemical looping combustion is a typical dual fluidized bed technology where chemically active bed material, i.e. metal oxide, is circulated between two fluidized bed reactors separated with respect to the gas phase. It has been initially proposed as early as 1954 as a method to produce pure CO_2 from oxidizable carbonaceous material [3]. Then, the concept was studied theoretically as a means to improve the reversibility of combustion [4-6]. More recently CLC has been identified as a promising technology for CO_2 capture from power plants, where 100 % of carbon capture can be achieved without a significant energy penalty [7-10]. In the so-called fuel reactor (FR), a hydrocarbon fuel is oxidized by oxygen released from the bed material. The global reaction in the FR can be summarized for the case of full oxidation as follows:

$$C_xH_y + \left(2x + \frac{y}{2}\right)MeO_\alpha \rightarrow xCO_2 + \frac{y}{2}H_2O + \left(2x + \frac{y}{2}\right)MeO_{\alpha-1} \quad (1)$$

This step takes place at temperatures between 750 and 1000 °C and is either slightly exothermic or even endothermic depending on the type of fuel and on the oxygen carrier chosen [11]. In the so-called air reactor (AR), the bed material is re-oxidized with air:

$$MeO_{\alpha-1} + \frac{1}{2}O_2 \rightarrow MeO_\alpha \quad (2)$$

This step is always strongly exothermic and the sum of reaction enthalpies of Reactions (1) and (2) is equal to the reaction enthalpy of direct combustion. Heat is withdrawn from the system with the hot exhaust gas streams. For common excess air ratios around 1.2, additional heat must be withdrawn directly from the reactors either by cooling the reactor walls or by cooling the circulating bed material stream. In practice, the extent in which fuel oxidation happens in the fuel reactor depends on the availability of oxygen (sufficient carrier circulation) and on the kinetics of the elementary gas-solid reactions (sufficient gas-solid contact time). If less oxygen is provided than needed for full oxidation, the chemical looping system operates in chemical looping reforming (CLR) mode. In this case the fuel reactor exhaust contains at least the combustible species CO and H_2 besides CO_2 and H_2O and the global heat release is decreased.

A key-issue for the industrial feasibility of CLC is the availability of oxygen carriers suitable from both the technical and the economical point of view. Therefore, extensive research is performed concerning the preparation and subsequent testing of materials as summarized e.g. by Johansson et al. [12].

The dual circulating fluidized bed (DCFB) system

For chemical looping combustion (CLC) or reforming (CLR) processes with selective oxygen transport by the bed material, the following basic requirements can be stated:

- High global solids circulation is required in order to provide enough oxygen in the fuel reactor and in order to keep the temperature difference between air reactor and fuel reactor low.
- Excellent gas-solids contact is required in both reactors in order to obtain satisfactory gas conversion. This is especially true for the fuel reactor of CLC systems where significant amounts of unconverted fuel will hardly be tolerable in applications.
- Low particle attrition rates are appreciated especially if costly oxygen carriers are to be used.

Therefore, the system investigated in the present study represents a combination of the existing practice in CLC technology, where a bubbling bed fuel reactor is added to the return leg of the solids of a circulating fluidized bed system, with the idea to use two scale-up ready CFBs for both air reactor and fuel reactor. A second aim of the present effort is to propose a robust fluidized bed system which is most simple with respect to necessary process control devices. In the dual circulating fluidized bed (DCFB) system according to Fig. 1, the two CFB reactors are interconnected via a fluidized loop seal in the bottom region of the reactors (lower loop seal). The entrainment of the left hand side reactor (primary reactor) determines global solids circulation. The solids are separated from the primary reactor exhaust stream in a cyclone separator and pass over through a fluidized loop seal (upper loop seal) into the right hand side reactor (secondary reactor). From there, the global solids loop closes via the lower loop seal. The secondary reactor features a circulation loop in itself (secondary reactor cyclone and internal loop seal) and may be optimized with respect to good gas-solid contact and low particle attrition rates. The global circulation rate can be effectively controlled by staged fluidization of the primary reactor. This is common practice in CFB technology. The direct hydraulic communication of the two CFB reactors allows stable solids distribution in the system as long as the lower loop seal is designed large enough to not significantly hinder solids flow. Imposing moderate pressure differences between the two reactors changes the theoretical solids levels in the system. This can be done by changing the backpressure from the exhaust gas lines and allows active control of the solids hold-up in each reactor.

Apart from CLC, the DCFB system has potential to be applied in other dual bed processes such as carbonate looping for end-of-pipe CO_2 capture [13], dual bed gasification and sorption enhanced reforming [14,15]. The design of the 120 kW fuel power pilot plant for CLC is described in the following.

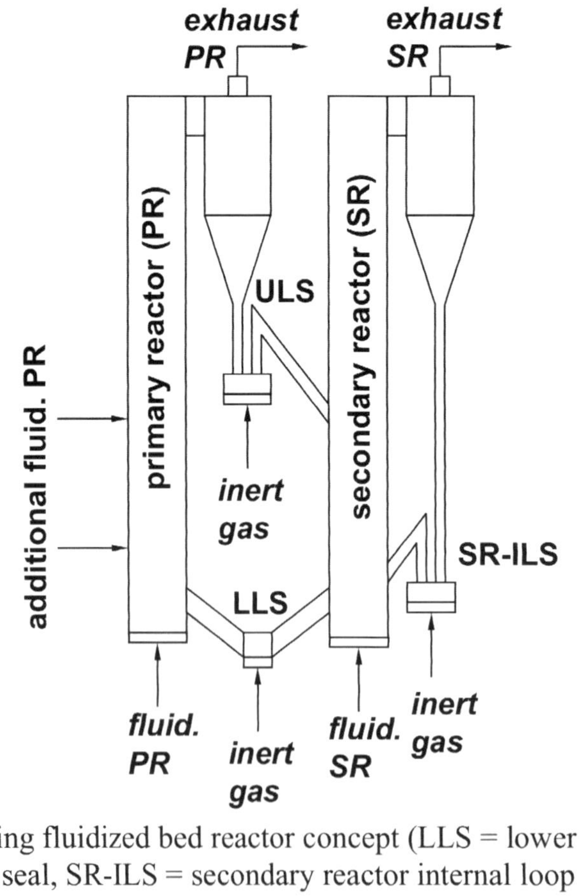

Figure 1. Dual circulating fluidized bed reactor concept (LLS = lower loop seal, ULS = upper loop seal, SR-ILS = secondary reactor internal loop seal)

Design of the 120 kW test rig

A DCFB test rig for CLC of gaseous fuels has been designed and built with the hot commissioning phase completed in early 2008. The air reactor (AR) of the CLC system is designed as the primary reactor in the sense of the DCFB concept described above. The fuel reactor (FR) is the secondary reactor, which may be optimized with respect to gas phase conversion. For means of simplicity at the small scale, the lower loop seal connecting the two reactors represents a continuation of the reactor bodies. The main fluidization nozzles are arranged along the circumference of the cylindrical reactor shells. The bulk geometry of the two reactors and the operating parameters of the design case are reported in Table 1.

In order to remove the heat released from combustion and to control the system temperature independently of the global air ratio, the air reactor shell is equipped with cooling jackets. These cooling jackets are operated either with boiling water or gaseous cooling media. The cyclone separators are designed according to Hugi [16]. The loop seals are fluidized with steam in nominal operation and may be switched to air fluidization during start-up and shut-down. The exhaust gas streams of the two reactors are cooled separately to about 573 K and analyzed on-line to evaluate the conversion of the fuel as well as the leakages of the loop seals. The cooled exhaust gas streams pass valves. These allow for imposing defined backpressure on each reactor. The exhaust gas streams are then mixed and sent to a natural gas fired post combustion unit, cooled again, filtered and sent to the chimney. Solids sampling is possible out of the upper and the lower loop seal during operation. This allows accurate interpretation of experimental results (carrier oxidation states and solids circulation rate).

Table 1. Basic geometry and design case operating parameters of the 120 kW CLC test rig.

Parameter	Unit	Air reactor	Fuel reactor
riser reactor free height	m	4.1	3.0
riser reactor inner diameter	mm	150	159
inlet gas flow	Nm^3/h	138.0	12.0
outlet gas flow	Nm^3/h	113.9	35.9
temperature in reactor	K	1213	1123
design fluid for calculation		depleted air	$H_2O/CO_2 = 2/1$
particle size	mm	0.12	
particle apparent density	kg/m^3	3200	
particle sphericity	-	0.99	
Archimedes number		7.55	9.13
superficial velocity	m/s	7.32	2.08
ratio U/U_{mf}	-	1280.4	315.4
ratio U/U_t	-	15.5	3.8
fuel power (natural gas)	kW	120	
lower heating value of fuel	MJ/kg	48.8	
design air/fuel ratio	-	1.2	

First operating results

Operation at chemical looping conditions started in January 2008 with natural ilmenite ($FeTiO_3$) as bed material. These tests, where hydrodynamics and syngas conversion have been studied, paved the way for operation with the designed NiO-based oxygen carrier which has been produced under the supervision of Chalmers University of Technology in Gothenburg/Sweden [17]. The first results for ilmenite as oxygen carrier in the 120 kW test rig has been recently reported elsewhere [18]. Here, in the context of the particularities of the DCFB reactor system, results from operation with NiO-based particles will be presented. Apart from steam for loop seal fluidization no steam or nitrogen is added to the fuel reactor.

Figure 2 shows the pressure profile in the reactor system for 150 kW fuel power with natural gas as fuel and at a global air ratio of 1.1. The primary reactor side (air reactor) shows a very distinctive exponential decay of solids with height. The solids profile in the secondary reactor (fuel reactor) show a steep decay of pressure in the lower part of the reactor. This indicates that the solids are concentrated in the dense zone at the bottom and the upper part of the fuel reactor is still relatively lean. The pressure differences in the loop seals represent pressure losses due to the flowing solids. From this point of view, the flow via the internal loop seal (i.e. the fuel reactor solids entrainment) is expected to be small compared to the global solids circulation between air reactor and fuel reactor.

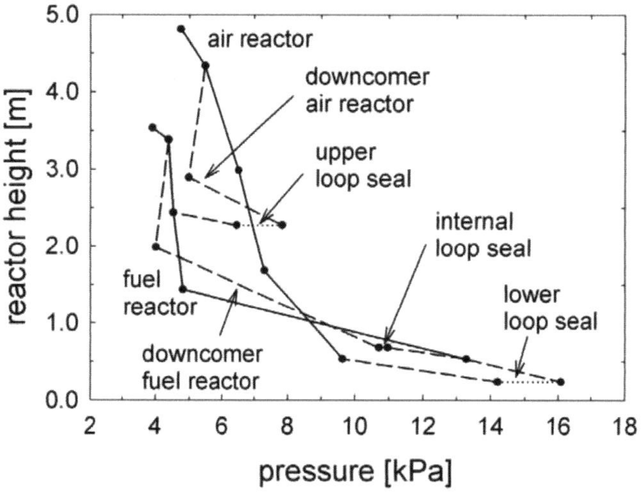

Figure 2. Pressure profile in the test rig at 150 kW fuel power and an air ratio of 1.1.

The most important performance criterion of CLC systems is the degree of fuel conversion in the fuel reactor. The critical output parameters methane conversion

$$X_{CH_4} = 1 - \frac{y_{CH_4}}{y_{CH_4} + y_{CO} + y_{CO_2}}, \tag{3}$$

and CO_2 yield

$$\gamma_{CO_2} = \frac{y_{CO_2}}{y_{CH_4} + y_{CO} + y_{CO_2}} \tag{4}$$

are shown Figure 3. The active solids mass in each reactor is obtained from pressure differences. It turns out that the CO_2 yield of the investigated oxygen carrier and plant design is strongly dependent on operating temperature. At lower temperatures the slip of unconverted fuel as CO and H_2 out of the fuel reactor is increased while the (primary) conversion of CH_4 seems to be less dependent on temperature. There is much room for further research especially with respect to the reaction mechanism and chemical kinetics of the fuel reactor. The CO_2 yield for propane operation also reported in Figure 3 indicates that the presence of higher hydrocarbons is not a problem in CLC. Contrarily, the CO_2 yield for C_3H_8 is even higher than for natural gas (CH_4). It must be realized that scaling up the results for the current pilot plant to larger scale conversion level will benefit from increasing reactor height from 3 m to e.g. 12-18 m typical for circulating fluidized beds.

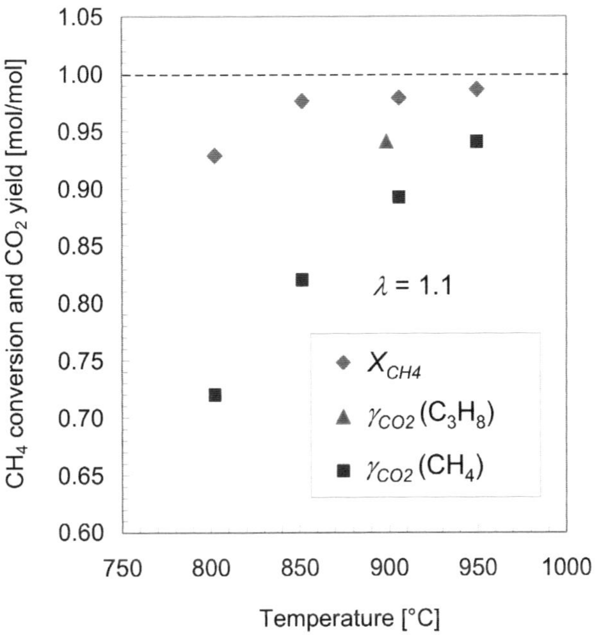

Figure 3. CH_4 conversion and CO_2 yield at the test rig as a function of fuel reactor operating temperature (global air ratio $\lambda = 1.1$, natural gas fuel power 145 kW, C_3H_8 fuel power 127 kW).

Conclusions

A system consisting of two circulating fluidized beds reactors hydraulically coupled by a loop seal in the bottom region of the two reactors is presented and investigated for chemical looping combustion using metal oxides for selective transport of oxygen. The proposed system is seen as scale-up ready candidate for large scale chemical looping application. The main characteristics are:

- high solids circulation rate
- low specific solids inventory
- stable solids distribution because of hydraulic connection between the two reactors
- secondary reactor flow regime can be optimized with respect to gas phase conversion
- global solids circulation can be controlled by staged fluidization of the primary reactor or by controlling the flow resistance in the lower loop seal
- solids hold up can be shifted between the reactors by imposing moderate backpressure from the exhaust gas lines

High solids circulation and low solids hold up result in a relatively narrow particle age distribution in each reactor.

The measured pressure profile indicates for the test rig that the circulating fluidized bed regime in the primary reactor is well developed with particles distributed along the whole height. On the secondary reactor side, however, the particles are concentrated in the bottom region and only a low particle flow is indicated for the secondary reactor's internal loop.

The fuel conversion has been found to be well above 90 % for the NiO-based oxygen carrier. For the investigated system, a temperature dependency of CO_2 yield has been found. A design with higher riser reactors is expected to significantly improve global fuel conversion in the dual circulating fluidized bed system for CLC.

Acknowledgement

This work was part of the EU financed project Chemical Looping Combustion CO_2 Ready Gas Power (CLC GAS POWER - FP6 Contract No. 019800), coordinated by Chalmers University of Technology. The project is also part of phase II of CCP (CO_2 Capture Project) via Shell.

Notation

Ar	Archimedes number	-
T	Temperature	K
U	superficial gas velocity	$m \cdot s^{-1}$
X_{CH4}	conversion of methane	$mol \cdot mol^{-1}$
y	gas phase mole fraction	$mol \cdot mol^{-1}$
γ_{CO2}	CO_2 yield based on carbon supplied	$mol \cdot mol^{-1}$

Super-/Subscripts

AR	air reactor
FR	fuel reactor
i	reference to gas species
mf	minimum fluidization velocity
t	terminal velocity for single particle

References

1. IEA (2006), "World Energy Outlook 2006," in *OECD Energy Series*, Priddle, R., ed.
2. Lyngfelt, A., Johansson, M., Mattisson, T (2008), "Chemical looping combustion – status of development," in *Circulating fluidized bed technology IX*, Werther, J. et al., eds. TuTech, Hamburg, Germany.
3. Lewis, W.K., Gilliland, E.R. (1954), "Production of pure carbon dioxide," U.S. Patent No. 2665972.

4. Knoche, K.F., Richter, H. (1968), "Verbesserung der Reversibilität von Verbrennungsprozessen," *Brennstoff-Wärme-Kraft* 20, pp.205-210 (written in German).
5. Richter, H.; Knoche, K.F. (1983), "Reversibility of combustion processes, efficiency and costing, second law analysis of processes," *ACS Symp. Series* 235, pp.71-85.
6. Ishida, M.; Zheng, D.; Akehata, T. (1987), "Evaluation of a chemical-looping-combustion power-generation system by graphic exergy analysis," *Energy* 12, pp.147-154.
7. Ishida, M., Jin, H., (1994), "A new advanced power-generation system using chemical-looping combustion," *Energy* 19, pp.415-422.
8. Ishida, M., Jin, H., (1997), "CO2 recovery in a power plant with chemical looping combustion," *Energy Conversion and Management* 38 (SUPPL. 1), pp.187-192.
9. Lyngfelt, A., Leckner, B., Mattisson, T. (2001), "A fluidized bed combustion process with inherent CO2 separation, application of chemical looping combustion," *Chemical Engineering Science* 56, pp.3101-3113.
10. Ryu, H.-J., Bae, D.-H., Jin, G.-T. (2002), "Chemical-Looping Combsution Process with Inherent CO2 Separation; Reaction Kinetics of Oxygen Carrier Particles and 50kWth Reactor Design," in *The World Congress of Korean and Korean Ethnic Scientists and Engineers*, Seoul, Korea, pp. 738-743.
11. Abad, A., García-Labiano, F., Adánez, J., de Diego, L.F., Gayán, P., Celaya, J. (2006), "Applicability limits of Cu-, Fe-, and Ni-based oxygen carriers in Chemical-Looping Combustion," in *8th International conference on greenhouse gas control technologies*, Trondheim, Norway.
12. Johansson, M., Mattisson, T., Rydén, M., Lyngfelt, A. (2006), "Carbon Capture via Chemical-Looping Combustion and Reforming," in *International seminar on carbon sequestration and climate change*, Rio de Janeiro, Brazil.
13. Abanades, J.C, Anthony, E.J., Lu, D.Y., Salvador, C., Alvarez, D., (2004) "Capture of CO2 from Combustion Gases in a Fluidized Bed of CaO," *AIChE J.* 50, pp.1614-1622.
14. Florin N.H., Harris, A.T. (2008), "Review of enhanced hydrogen production from biomass with in situ carbon dioxide capture using calcium oxide sorbents," Chem. Eng. Sci. 63, pp. 287-316.
15. Pröll, T., Hofbauer, H. (2008), "H2 rich syngas by selective CO2 removal from biomass gasification in a dual fluidized bed system — Process modelling approach," *Fuel Processing Technology*, In Press, doi:10.1016/j.fuproc.2008.05.020.
16. Hugi E. (1997), "Auslegung hochbeladener Zyklonabscheider für zirkulierende Gas/Feststoff-Wirbelschicht-Reaktorsysteme," *Fortschr.-Ber. VDI* 3 (502), VDI, Düsseldorf, Germany (written in German).
17. Johansson, M., Mattisson, T., Lyngfelt, A. (2006), "Use of NiO/NiAl2O4 particles in a 10 kW chemical-looping combustor," *Ind. Eng. Chem. Res.* 45, pp.5911 -5919.
18. Pröll, T., Mattisson, T., Mayer, K., Bolhar-Nordenkampf, J., Kolbitsch, P., Lyngfelt, A., Hofbauer, H. (2008), "Natural minerals as oxygen carriers for chemical looping combustion in a dual circulating fluidized bed system," in Proceedings of the 9th International Conference on Greenhouse Gas Technologies, Washington DC.
19. Kolbitsch P., Pröll, T., Bolhar-Nordenkampf, J., Hofbauer, H. (2008), "Characterization of chemical looping combustion pilot plant performance via experimental determination of solids conversion," submitted to *Energy and Fuels*.

Coal Direct Chemical Looping (CDCL) Process

Fanxing Li, Hyung Kim, Deepak Sridhar, Liang Zeng, Fei Wang, L. -S. Fan[1]
The Ohio State University, Columbus, OH. U.S.A.

Abstract

The Coal Direct Chemical Looping (CDCL) process is a novel technology that produces hydrogen from coal with in-situ CO_2 separation. The CDCL uses a composite iron oxide particle as an oxygen carrier for coal conversion. These particles are reduced by coal to capture CO_2 in the reducer. The reduced particles then are used in the second reactor to produce hydrogen through oxidation using steam. They can also be used to react with air for heat and power generation. The indirect combustion of coal with composite iron particles significantly enhances the efficiency of the hydrogen production process. The demonstration results carried out in a bench scale unit show more than 95% conversion of coal and more than 99% conversion of coal volatiles.

In this paper, the CDCL process is first reviewed. The demonstration results are then presented. Effect of sulfur and ash on the performance of the composite iron oxide particle is also discussed.

1. Introduction

Coal combustion power plant accounts for a third of the total CO_2 emission in the United States (Herzog, 1997). In order to capture and sequestrate the CO_2 from the coal combustion flue gas, significant energy and economic penalty have to be paid. For instance, CO_2 capture can reduce the energy conversion efficiency of a coal combustion power plant by up to 40% (NETL, 2007, Li, 2008). With projected increase in coal consumption for power generation, hydrogen generation, and fuel synthesis, novel coal conversion techniques that can reduce the CO_2 capture cost with improved energy conversion efficiency is highly desirable. Chemical-looping combustion, which integrates CO_2 capture in the fuel conversion scheme, is a promising approach for clean and efficient coal conversions (Ishida, 1994 and Mattisson, 2003).

In this paper, a novel coal direct chemical looping (CDCL) process which directly converts coal into hydrogen and/or electricity is presented. Experimental results obtained from a bench scale moving bed unit is reported and discussed.

2. Process Overview

The CDCL process, illustrated in Figure 1, utilizes composite Fe_2O_3 particles to convert coal into hydrogen and/or electricity at any relative proportions (Gupta, Velazquez-Vargas et al. 2004; Thomas, Fan et al. 2004; Fan, Gupta et al. 2007). In the CDCL process, composite Fe_2O_3 particles are introduced into the reducer along with pulverized coal and sub-stoichmetric amount of oxygen from the Air Separation Unit (ASU). With a desired gas-solid contacting pattern, coal is gasified in-situ and reacted

---

[1] Corresponding Author

with Fe_2O_3 particles. As a result, a mixture of metallic iron and Wuestite is produced along with a flue gas stream composed of CO_2, H_2O, and contaminants such as H_2S and elemental mercury. After condensing out the steam, the flue gas can be compressed and sequestrated. The reduced metallic iron/Wuestite particle from the reducer will then enter the oxidizer to react with steam to form hydrogen. The resulting Fe_3O_4 exiting the oxidizer will be subsequently combusted with air in the entrained flow combustor, which conveys the particle back to the reducer pneumatically while regenerating the particle to its original fully oxidized form. The heat released in the combustor will be partially carried by the exhaust gas, which can be used for steam or electricity generation.

Among the three main reactors involved in the CDCL process, the reducer is the most challenging since it involves complex solid-solid reactions scheme between the Fe_2O_3 composite particles and pulverized coal. Therefore, the process testing revolves around the validation of the concept of the reducer.

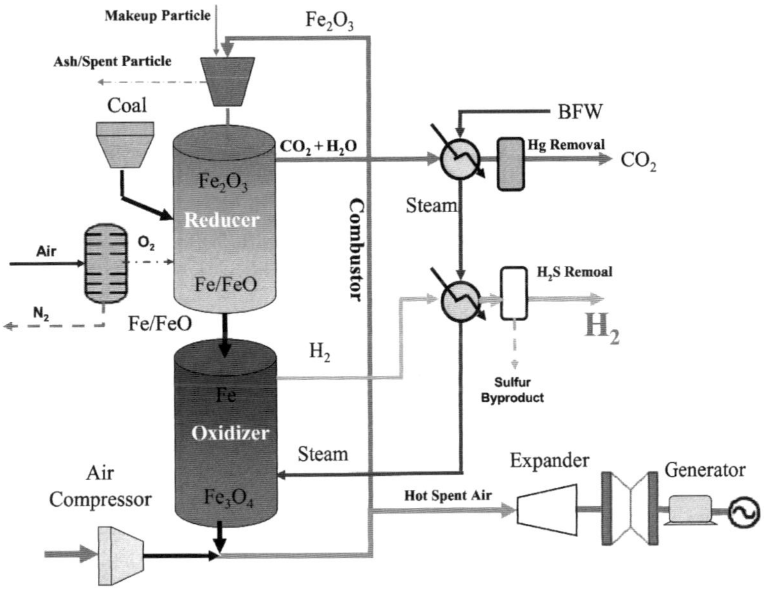

Figure 1 Schematic diagram of Coal-Direct Chemical Looping process

3. Experimental Setup

3.1 Reactor Setup

A bench scale demonstration unit for CDCL demonstration was designed and constructed at the Ohio State University. Figure 2 illustrates the schematic diagram of the unit. The unit was designed to handle solid flow rates up to 1.36 g/s (4.9 kg/hr) with gas flow rates up to 200 ml/s (12 l/min). The reacting section of the reactor has an I.D. of 1.6 inches and a height of 40 inches. In a typical demonstration, solid reactants such as the oxygen carrier particle and/or coal char are first loaded to the top funnel and then moved downwards steadily by the screw conveyor system. The gas composition along the axial positions of the reactor is constantly monitored using a Varian CP-4900 micro GC during the test and the solid can be taken from the solid sampling ports for further characterization after the experiment. Detailed information regarding the reactor setup is described in Gupta et al. (Gupta, Velazquez-Vargas et al. 2007)

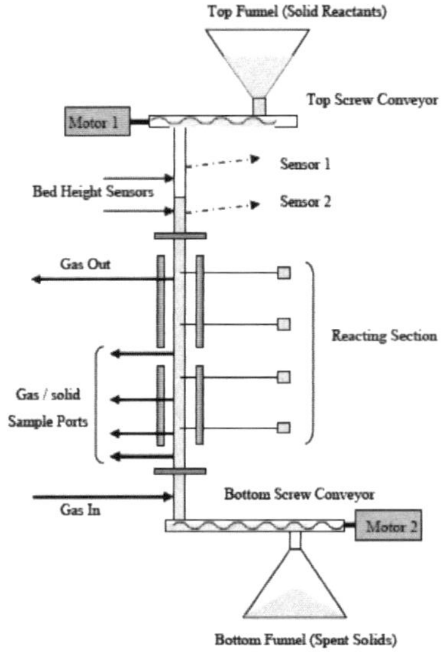

Figure 2. Schematic diagram of bench scale reactor

3.2 Experimental Procedures

Two experiments were carried out to demonstrate the feasibility of the proposed reducer configuration. The first experiment validates the conversion of methane, which represents one of the most stable form of coal volatile, using Fe_2O_3 composite particles. The second experiment validates the conversion of coal char with the presence of oxygen and Fe_2O_3 composite particles.

Before the methane experiment, the Fe_2O_3 composite particles were loaded in the top funnel. The particles were then moved downwards steadily by the screw conveyor system at a preset flow rate which is 31.5 g/min. Reactant gases were then introduced from the gas inlet located at the bottom of reactor. In the char conversion experiment, coal char and Fe_2O_3 particle mixture was loaded in the top funnel with oxygen introduced at a gas inlet located at the middle section of the reactor.

After the moving bed operation, the solid samples were taken from the various sampling ports for the further analysis using a carbon analyzer (CA) (UIC) and a thermogravimetic analyzer (TGA) (Perkin Elmer, Pyris 1). The carbon analyzer determines the carbon deposition and coal char conversion while the TGA verifies the conversion of particles through the weight change of the particle during air oxidation.

4. Results and Discussions

Figure 3 shows the gas conversions (red dotted line) and solid conversions (black solid line) at different ports.

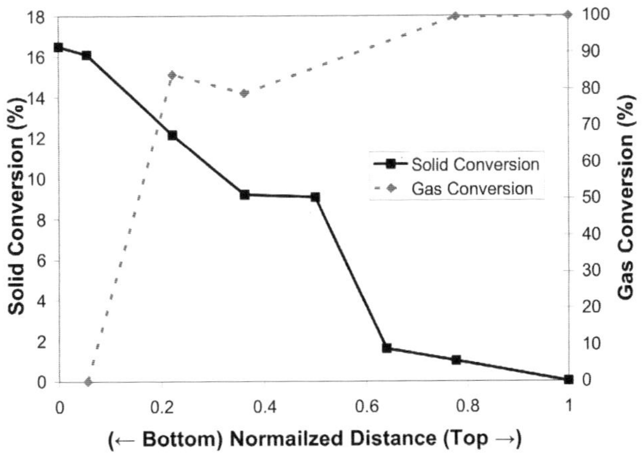

Figure 3. Gas and solid conversions at various solid sampling ports

As it is seen in Figure 3, around 16.5 % the solid conversion was achieved at the solid outlet whereas the upper parts of reactor observed less than 2% solid conversion. The solid conversion gradually increased as it approaches the bottom of the reactor. However, a methane conversion of nearly 100 % (99.89 %) was achieved at the gas outlet of the reactor, which corresponds to higher than 99% CO_2 concentration at the reactor outlet (dry basis). Figure 4 shows the gas concentrations at various gas sampling ports.

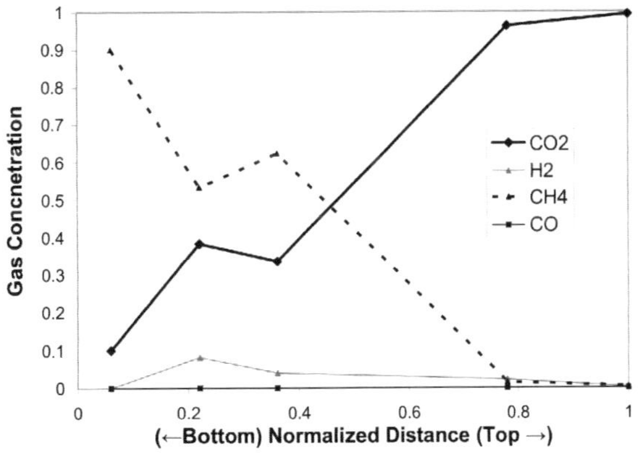

Figure 4. Gas concentrations at various gas sampling ports

The oxidation of coal char using the composite particles was also performed in the bench scale moving bed unit. The coal char obtained from bituminous (Pittsburgh #8) has been tested showing over 95% conversion as shown. The coal char conversion according to the normalized distance from the particle/coal mixture inlet is shown in Figure 5. Solid residence time of ~ 80 minutes was used.

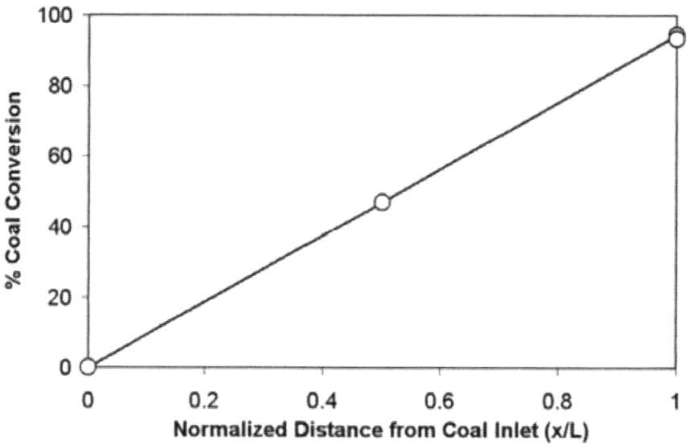

Figure 5. Coal char conversion in the presence of iron oxide composite particle and oxygen

The reduced Fe_2O_3 composite particles were further tested in an X-Ray Diffraction instrument (Scintag XDS2000) to determine the fate of sulfur. It was found that FeS was formed during the reducer operation. Further tests on the reactivity of the particle showed that the particle was fully recyclable after reacting with coal. More than 70% of the ash in coal was able to be separated from the particle using mechanical sieving method.

5. Conclusions

A novel coal direct chemical looping process that converts coal into hydrogen and/or electricity is discussed. Bench scale experiments show that the proposed reducer reactor is capable to convert around 99.9% coal volatiles and 95% bituminous coal char while reducing the Fe_2O_3 composite particles. At least a portion of the sulfur in coal was found to react with the Fe_2O_3 composite particle to form FeS. Moreover, ash can be separated from the particle using mechanical method. It is further determined that the Fe_2O_3 particle can be regenerated with air without losing its reactivity after reacting with coal. To conclude, the CDCO process concept is feasible according to preliminary testing results obtained from a bench scale units. Further improvements in coal char conversion, ash separation, and reduction in solid residence time are desirable.

6. Acknowledgement

This work has been supported by the Ohio Coal Development Office of the Ohio Air Quality Development Authority, U. S. Department of Energy, Ohio State University and an industrial consortium.

7. References

1. Fan, L.-S., P. Gupta, et al. (2007). Systems and methods of converting fuels. *PCT International Applications* WO 2007082089.
2. Gupta, P., L. G. Velazquez-Vargas, et al. (2004). Hydrogen production from combustion looping (solids-coal). *Proceedings of Clear Water Coal Conference.*
3. Gupta, P., L. G. Velazquez-Vargas, et al. (2007). "Moving bed reactor setup to study complex gas-solid reactions." *Review of Scientific Instruments* **78**(8)
4. Herzog, H.; Drake, E.; Adams, E. CO_2 Capture, Reuse, and Storage Technologies for Mitigating Global Climate Change. A White Paper; *U.S. Department of Energy*: Washington, D.C., January, 1997, U.S.DOE/DEAF22- 96PC01257
5. Ishida, M.; Jin, H. G., A New Advanced Power-Generation System Using Chemical-Looping Combustion. *Energy* (1994), 19, (4), 415-422.
6. Mattisson, T.; Jarders, A.; Lyngfelt, A. Reactivity of Some Metal Oxide Support on Alumina with Alternating Methane and Oxygen Application for Chemical-Looping Combustion. *Energy Fuels* 2003, 17, 643-651.
7. Thomas, T., L.-S. Fan, et al. (2004). Combustion looping using composite oxugen carriers. *US Patent Applications*. U.S. 11,010,648

686f Calcium Looping Process for Hydrogen, Electricity and Liquid Fuel Production with In-Situ Carbon Dioxide and Sulfur Capture

Shwetha Ramkumar, Mahesh Iyer, Zhenchao Sun, and Liang-Shih Fan. Chemical and Biomolecular Engineering, The Ohio State University, 140 W. 19th Ave, 125A Koffolt Labs, Columbus, OH 43210

The Calcium Looping Process is one of the clean coal technologies being developed for the production of hydrogen, electricity and liquid fuels from syngas obtained via coal gasification. It integrates the water gas shift (WGS) reaction with in-situ carbon dioxide (CO2), sulfur and halide removal at high temperatures in a single reactor while eliminating the need for a water gas shift catalyst thus reducing the overall foot print of the hydrogen production process. The calcium looping process comprises of two reactors: the carbonation reactor where the thermodynamic constraint of the water–gas-shift reaction is overcome by the incessant removal of the CO2 product from the reaction mixture and high-purity hydrogen is produced with contaminant removal and the calciner where the calcium sorbent is regenerated and a sequestration-ready CO2 stream is produced.

Process evaluation and optimization of the calcium looping system for the production of hydrogen from coal has been conducted using the commercial ASPEN Plus® process simulation software. The over all efficiency of this process for the production of 99.999% pure hydrogen from coal is 63% (HHV) when compared to the state-of-the-art, hydrogen from coal process which has an efficiency of 57% (HHV). The purity of hydrogen is increased by a large extent when the carbonation reaction is integrated with the WGSR. The steam addition for the WGSR can also be reduced to near-stoichiometric quantities which aids in reducing the parasitic energy consumption of the process. In addition, the extent of H2S removal by the CaO sorbent is also enhanced by operating at lower steam concentrations. High purity hydrogen of 99.7% with less that 1 ppm sulfur impurity has been obtained in a bench scale fixed bed reactor system. Experiments conducted in a bench scale system have revealed the superior performance of the OSU patented PCC based calcium oxide sorbent for the production of hydrogen.

A Comparative Study of the Gasification Technology *versus* Traditional Combustion Technologies for Power Generation in Mexico

Emilio Manzanares-Papayanopoulos*, Manuel Fernández-Montiel, Jorge A. Altamirano-Bedolla, Alejandro Mani-González, and Alejandro Arriola-Medellín

*Gerencia de Procesos Térmicos, Instituto de Investigaciones Eléctricas, Reforma No. 113 Col. Palmira, CP62490, Cuernavaca, Morelos, México. *e-mail: emp@iie.org.mx*

ABSTRACT

A study of the gasification technology versus traditional combustion technologies for power generation was performed in order to present the cost-benefit of them, also, to verify the applicability of the gasification technology in the power generation in Mexico.

At the moment, the gasification technology is not applied in the power generation in Mexico. However, seen the importance of its worldwide application, there has been an increasing interest from the main power energy company, Comision Federal de Electricidad (CFE) and the main oil company, Petroleos Mexicanos (PEMEX) in the potential use that the gasification technology offers as an alternative for power generation. In addition, the technology allows synergies to other industries, the diversification in the use of different fuels, ultra low contaminant emissions, the used of the capture and sequestration of CO_2 technology, among others.

It is well known that the gasification technology as Integrated Gasification Combined Cycle (IGCC) become an attractive technology since it offers a clean path to power generation and flexibility on the feedstock, giving as a result a wide fuel diversification to be used in power plants.

The combustion technologies considered on this study were, subcritical pulverized coal (PC) and heavy fuel oil (HFO) steam plants, supercritical pulverized coal (SPC) steam plants, circulating fluidised bed (CFB) steam plants and natural gas combined cycle (NGCC) plants. The simulation of the power plants on this study was performed by means of Thermoflow Inc. software. In the case of IGCC plants, different feedstock, such as coal, petroleum coke and heavy fuel oil were included in the evaluation.

The comparative study was performed on the base of a techno-economic evaluation in function of the operative parameters and technical factors such as plant location, power energy efficiency, requirement of water process, emissions, etc. and, economical factors such as investment, fuel costs, etc. The results have shown that the application of the IGCC technology can be considered as an attractive alternative for power generation in Mexico, since it offers several advantages over the traditional combustion technologies.

Keywords: Electric power generation, IGCC, Gasification, Traditional combustion technologies, Fossil fuels, Efficiency, Costs

INTRODUCTION

It is well known that the electricity demand have been growing around the world. In Mexico, that situation is not an exception, and as a result, there is a permanent scheme for an expansion in the current power plant fleet (SENER, 2006).

However, the volatility prices of natural gas and crude, also the worldwide needs to reduce the greenhouse gas emissions point out a clear path to explore alternative technologies for power generation.

In past years, the Instituto de Investigaciones Electricas (IIE), considered as the technological research arm of the Energy sector in Mexico, has performed several studies of the gasification technology as Integrated Gasification Combined Cycle (IGCC) for power generation (Manzanares-Papayanopoulos, 2006; Fernández-Montiel, *et. al.*, 2003; Fernández-Montiel, 1994; Fernández-Montiel and Alanis, 1993 and 1994).

In a previous paper (Manzanares-Papayanopoulos *et. al.*, 2006), the study of the syngas production analysis as a function of fuel compositions, temperatures and pressures by simulation carried out by commercial computational packages, and the comparison of the energy produced as function of the fuel composition at the same operational parameters was presented. Additionally, a design of an atmospheric pressure gasification system at work-bench scale to gather information of the kinetics gasification process has been also presented (Altamirano-Bedolla *et. al.*, 2007).

The IGCC offers flexibility of the fuel used and with a required pollution reduction systems allows clean power generation (Higman and Der Burgt, 2003; Rezaiyan and Cheremisinoff, 2005).

There are other comparative studies of the gasification technology. Zheng and Furinsky (2005) have presented a comparative study in function of the features of four gasifiers used in an IGCC plant scheme. In other hand, Beér (2007) reviewed different electric power generation systems as a function of the plant efficiency; and Ordorica-Garcia *et al.* (2006) have presented a technical and economic comparison of the performance simulation of four IGCC plant designs with different levels of CO_2 capture against a NGCC plant without CO_2 capture.

In this work, the comparison of the IGCC performance against the most common traditional combustion technologies for power generation in function of the feedstock available in the Mexico is presented.

CASE DESCRIPTION

The traditional combustion plants for power generation considered for this study were two subcritical steam plants fed with pulverized coal (PC) and with heavy fuel oil (HFO), one supercritical pulverized coal (SPC) steam plant, two circulating fluidized bed (CFB) combustor plants fed with PC and with petroleum coke (petcoke), and a natural gas combined cycle (NGCC) plant.

The site conditions were also considered, since the efficiency of the gas turbines are affected due to the altitude. The technical evaluation analysis was performed for two feasible sites in the country, on-shore and in the mainland. The site conditions are presented in Table 1. The analysis of the feedstock available in function of the technology studied for each site is presented in Table 2.

Table 1 Site ambient conditions for the technical evaluation of the power generation plants

Site	Maximum Ambient Temperature	Design Ambient Temperature	Altitude	Atmospheric Pressure	Relative Humidity	Wet Bulb Temperature
Units	**(°C)**	**(°C)**	**(m)**	**(bar)**	**(%)**	**(°C)**
Mainland	36	33	1,589	0.844	32.62	20
On-shore	41.0	37	10	1.012	42.25	26

Table 2 Feedstock available in function of the site and the technology of the power generation plants

Site	Available Fuel	Technology
Mainland	Natural Gas	Combined cycle
	Petroleum coke	Fluidized bed
	Heavy fuel oil	Subcritical steam plant
On-shore	Natural Gas	Combined cycle
	Petroleum coke	Fluidized bed
	Heavy fuel oil	Subcritical steam plant
	Coal	Subcritical steam plant Fluidized bed Supercritical steam plant

The energy production capacity of the plants was of 700 MW (+/- 15%). The steam turbine cooling for the plants situated mainland was via aero-condensers, and for the plants situated on-shore was via sea water.

In reference of the premises for the economical evaluation, these are described as follow. The equipment cost of the plants was obtained directly for the Thermoflow software programs used in the technical and economical evaluation. The evaluation assumes a plant life of 30 years, a rate of return of 12%, a load factor of 80% to fix the comparative basis for the IGCC with the others technologies.

Figure 1 shows the price scenario of the feedstock for 30 years used in the economic evaluation of the technologies. The dotted line is the critical line of the fuel cost used in a IGCC plant in function of the cost of a NGCC plant (Jones and Shilling, 2002), i.e. below the dotted line any fuel costs make the IGCC plant economical attractive and above the dotted line the IGCC is not economical attractive. Coal 1 is from Colombia and Coal 2 is from Australia. The heavy fuel oils and the coke are from the Mexican refineries.

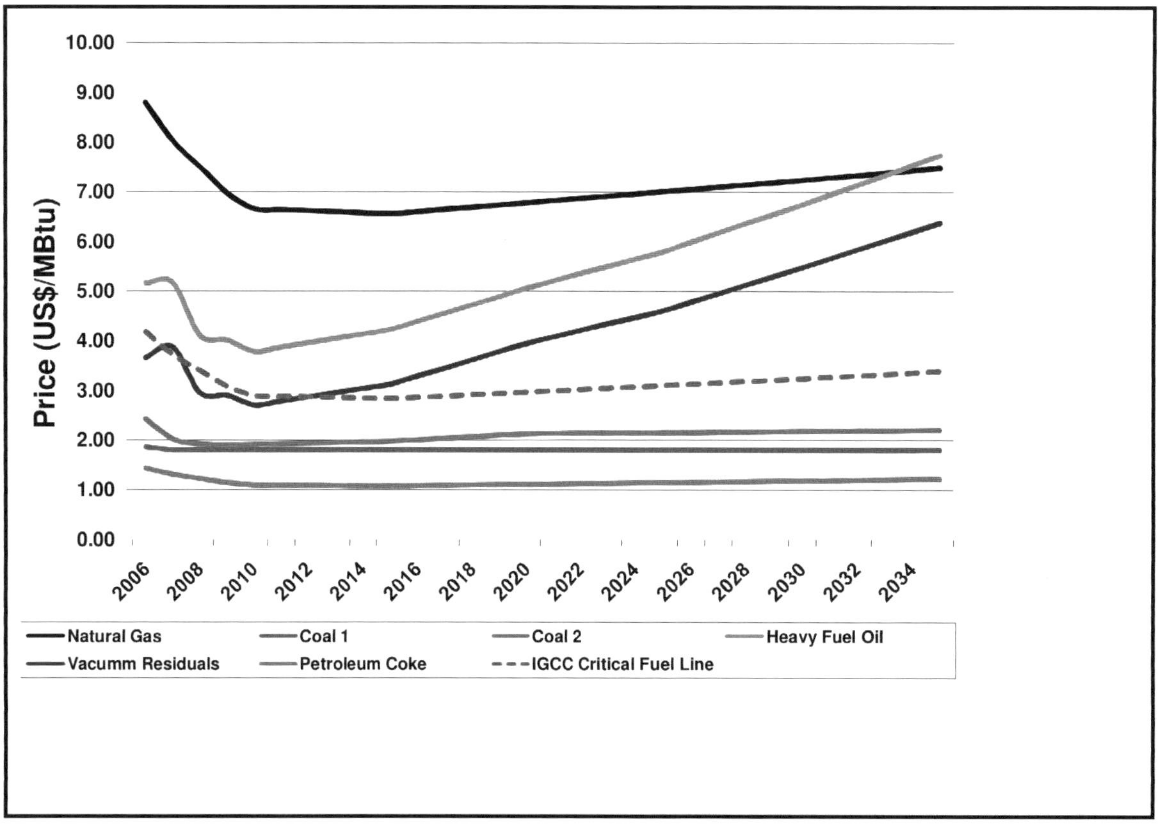

Figure 1 Price scenario to up 30 years of the feedstock available for the power generation plants in Mexico

RESULTS

The analysis of each technology was carried out in base of the plant performance. In function of that performance, a capital cost (initial investment) was associated for each plant. Figure 2 and Figure 3 shown the capital cost for the plants situated mainland and on-shore, respectively.

In both Figures, the lower investment cost is for a NGCC plant. However, the following attractive option in mainland for investment becomes the IGCC plant operated with heavy fuel oil. However, the first attractive IGCC plant for investment on-shore is until the fourth option presented in Figure 3.

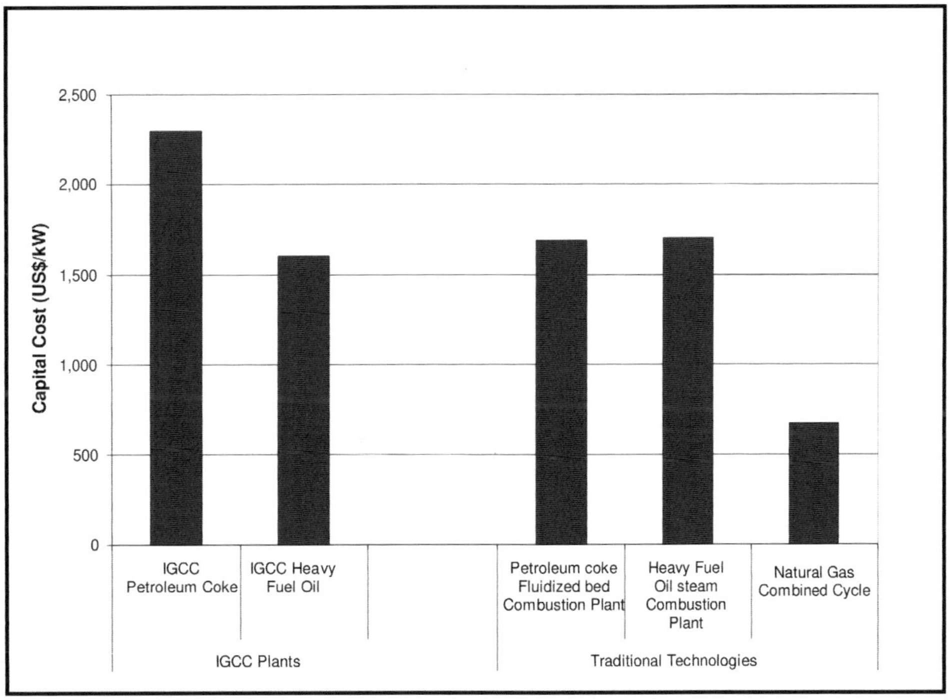

Figure 2 Direct capital costs of IGCC plants and traditional combustion plants in mainland

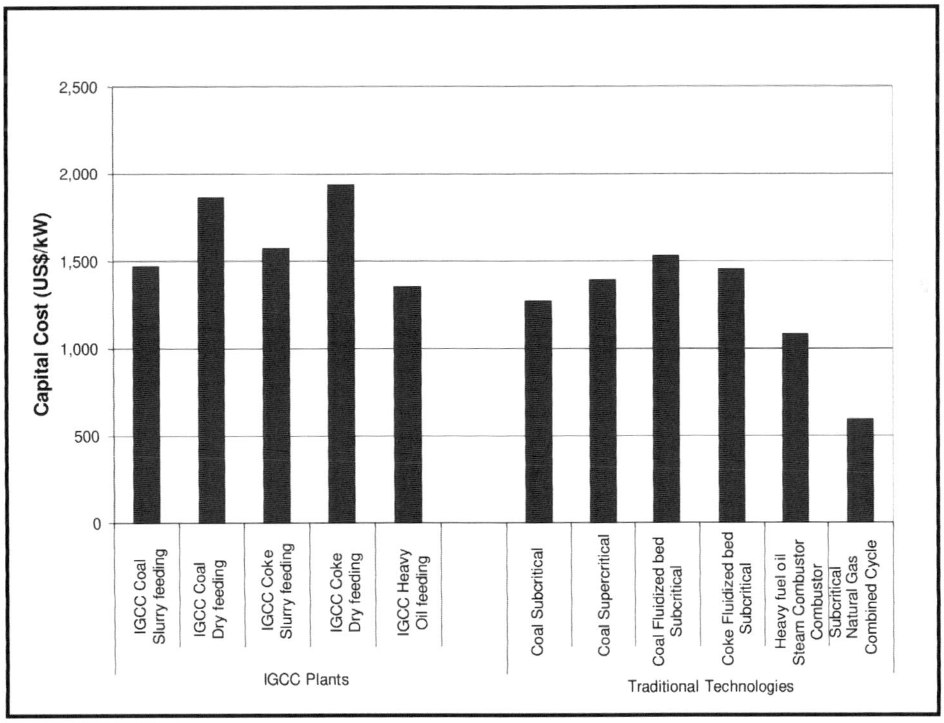

Figure 3 Direct capital costs of IGCC plants and traditional combustion plants on-shore

An interesting outcome on these Figures, it is that using the petroleum coke as feedstock results the highest inversion required to build an IGCC power generation plant. Since the petcoke from Mexican refineries has high levels of sulphur content, the cleaning systems needed to remove the sulphur are bigger.

A more deep analysis had required to include the indirect costs, i.e. power generating costs. Figures 4 and 5 present a comparative plot of the required investment costs including the operative costs (fuel, maintenance and operation costs) of each technology for the plants situated mainland and on-shore, respectively

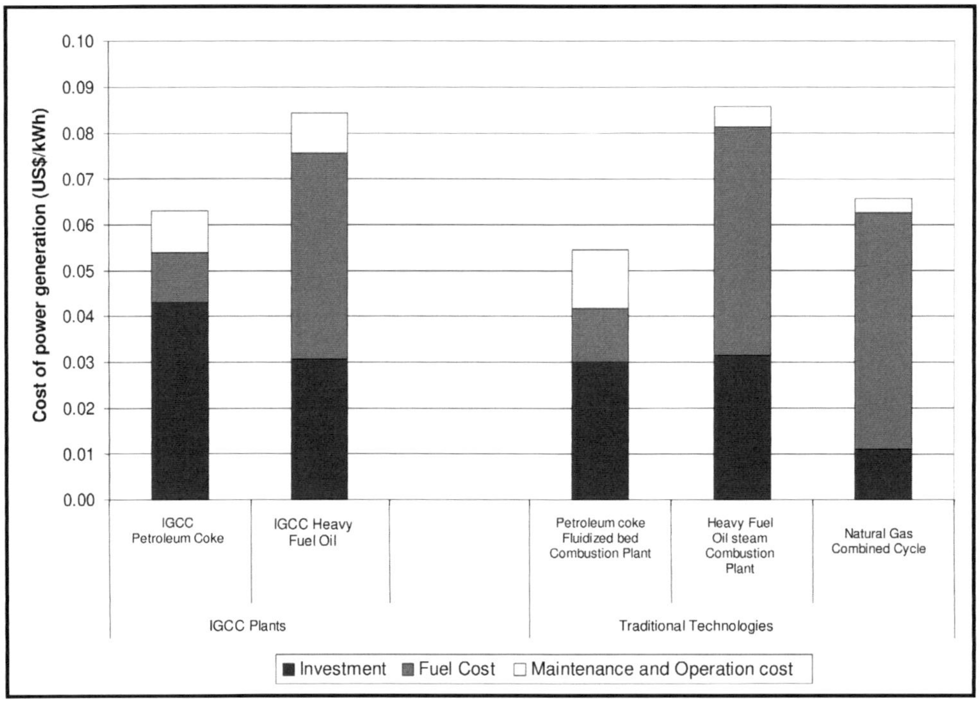

Figure 4 Comparative graph of the required investment costs including the operative costs of each technology for the plants situated mainland

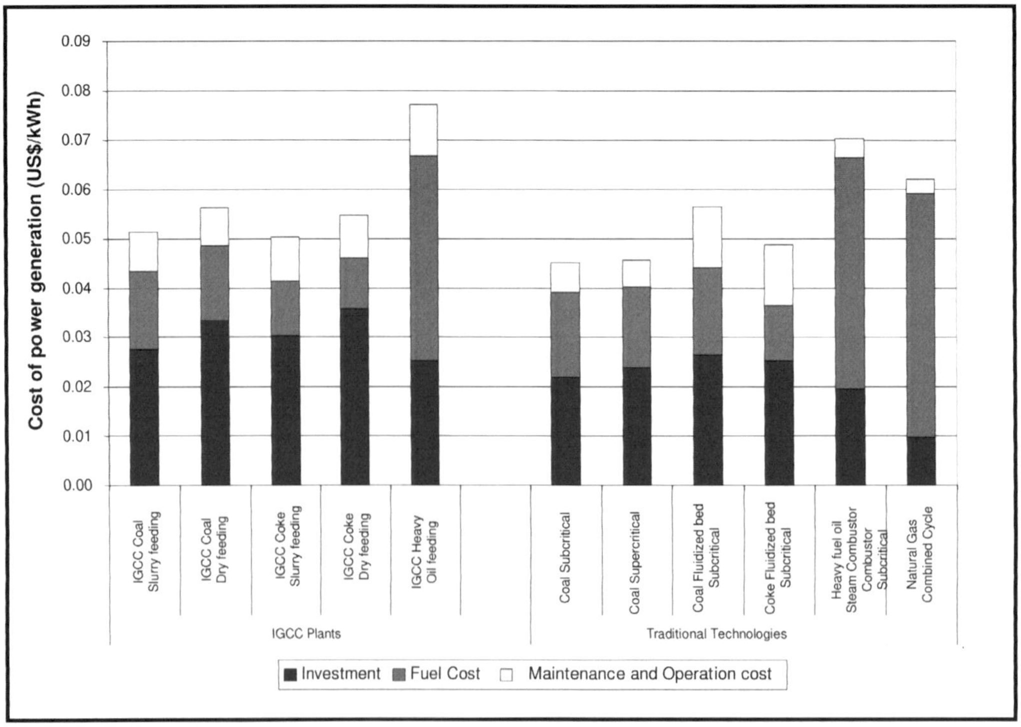

Figure 5 Comparative graph of the required investment costs including the operative costs of each technology for the plants situated on-shore

The outcome of the plant cost including the indirect costs pointed out that the analysis must be carry out including all the costs involved in the construction, maintenance and operation of a power generation plant. In the case of the plants situated mainland, see Figure 4, the option more attractive is the circulating fluidized bed combustion plant using as feedstock the petroleum coke. As a second option become the IGCC plant using also as feedstock the petroleum coke. The NGCC plant was the first option for just the investment cost point of view. However, due to the actual cost of the natural gas and the increasing cost scenario, the NGCC plant becomes as a third attractive option.

A same situation results for the analysis of the power generation plants on-shore, as shown in Figure 5. Four of five IGCC options are more attractive than the natural gas combined cycle plant. However, coal subcritical and supercritical steam generation plants are the most attractive options on-shore for power generation investment.

CONCLUSIONS

There are many conclusions from the work presented here. The technical evaluation pointed out that for a 700 MW IGCC plants three independent lines of syngas production are needed, and to have higher operational flexibility should be better do not have an air integrated system. Also, if the disponibility of an IGCC plants want to be higher then the plant most considered a spare gasifier. However, this action will increase the capital cost about 8.3%. In reference to the emissions, the IGCC plants are good option since they have similar emissions than a natural gas combined cycle plants without take care of the nature of the feedstock used.

The combination of the technical and the economical evaluation of the IGCC plants are presented next. The best option for an IGCC plant is the one presented on shore using petcoke slurry as feedstock, due to the low cost of the petroleum coke. Nonetheless, the IGCC plants are attractive in both sites on-shore and mainland, the pulverized coal subcritical and supercritical steam combustion plants are the best options for an investment. This is due to the evaluation of these coal plants was performed without desulphuration equipment since the emissions are inside of the environment normativity.

The outstanding result here is that the IGCC plants are competitive options against the natural gas combined cycle, and then it become important since the IGCC technology has the flexibility of the used different kind of fuels. However, in spite of the advantages of the IGCC technology, the natural gas combined cycle plants are more attractive since the capital cost will return in less time than the capital cost of the IGCC plants.

Finally, as an strategy of reducing costs, the Gasification Technology should be considered to be used in different synergies with other industries such as the oil industry, since there are different industrial processes than can take advantages of the subproducts obtained from this technology or to use the syngas as a feedstock to other industrial process as the Fischer-Tropsch process to obtain diesel and gasoline.

ACKNOWLEDGEMENT

The authors wish to thank to the Comision Federal de Electricidad (CFE) and to the Consejo Nacional the Ciencia y Tecnologia (CONACyT) for the financial support of this project.

REFERENCES

Altamirano-Bedolla J.A., Mani-González A., Manzanares-Papayanopoulos E., Fernández-Montiel M., and Romo-Millares C.A. (2007), "Preliminary Study of the Gasification Process of Mexican Fuels at Atmospheric Pressure", AIChE Spring Meeting 2007, Paper 9e, April, 2007, Houston, TX., USA.

Beér, J.M.; (2007), "High efficiency electric power generation: The environmental role", Progress in Energy and Combustion Science Journal, Vol. 33, pp 107-134.

Fernández-Montiel, M., Alcaraz-Calderón, A. M. y García-Chávez, L.; (2003), "Estudios y Desarrollo de Alternativas de Gasificación para el Sector Eléctrico", Informe 12585, Instituto de Investigaciones Eléctricas, Diciembre 2003, México.

Fernández-Montiel, M. y Alanís, F.J.; (1994), "Estudios de Factibilidad para Plantas de Gasificación de Carbón y de Combustóleo Integradas a Ciclos Combinados", Informe 44288, Instituto de Investigaciones Eléctricas, Marzo 1994, México.

Fernández-Montiel, M; (1994), "Reporte de actualización sobre el estado que guardan las tecnologías de gasificación aplicadas a la generación de energía eléctrica", Informe 44256, Instituto de Investigaciones Eléctricas, Octubre 1994, México.

Fernández-Montiel, M. y Alanís, F.J. (1993), "Reporte del Estado Actual de las Tecnologías de Gasificación Aplicadas a la Generación de Energía Eléctrica", Informe 44207, Instituto de Investigaciones Eléctricas, Octubre 1993, México.

Higman C. and Der Burgt M. Van, (2003), "Gasification", Elsevier.

Jones, R.M.; Shilling, N.Z.; (2002), "IGCC gas turbines for refinery applications", presented at 2002 Gasification Technologies Conference, San Francisco, Ca., USA.

Manzanares-Papayanopoulos, E., Fernández-Montiel, M., Altamirano-Bedolla, J.A., Alcaraz-Calderón, A.M., and Romo-Millares, C.A., (2006), "Development and Evaluation of the Gasification Technology as Alternative Power Generation", AIChE Spring Meeting 2006, Paper 128e, April, 2006, Orlando, Fl., USA.

Ordorica-Garcia, G.; Douglas, P.; Croiset, E.; Zheng, L.; (2006), "Technoeconomic evaluation of IGCC power plants for CO2 avoidance", Energy Conversion and Management Journal, Vol. 47, pp. 2250-2259.

Rezaiyan J. and Cheremisinoff, N. P.; (2005), "Gasification Technologies: A Primer for Engeneers and Scientists", Taylor & Francis.

SENER (Secretaría de Energía) (2006), "Prospectiva del Sector Eléctrico 2006-2015", Dirección General de Planeación Energética, Primera Edición.

Zheng, L. and Furinsky, E. (2005), "Comparison of Shell, Texaco, BGL and KRW gasifiers as part of IGCC plant computer simulations", Energy Conversion and Management Journal, Vol. 46, pp. 1767-1779.

CFD Modelling Gas-Solid Flows in CFB/FCC Riser Reactors: Simulation Using Kinetic Theory of Granular Flow (KTGF) in a Fully Developed Flow Situation

M. N. Idris[1, 2] and **A. Burn**[3]
[1]School of Process Environmental and Material Engineering
University of Leeds, Leeds LS2 9JT, United Kingdom
[2]Dept of Chemical Engineering, University of Maiduguri, Nigeria
[*]Tel: +44-113-2160568, Email: chemni@leeds.ac.uk
[3]Multiphase Flow Programming Specialist
ANSYS UK Limited
United Kingdom

Abstract

This study presents the simulation of gas-solid hydrodynamics and flow behaviour in a riser in an FCC riser circulating fluidised bed (CFB) reactor using the Eulerian-Eulerian two-fluid modelling approach, incorporating a kinetic theory constitutive model for dilute assemblies of FCC particulate solid. The interaction between gas and particles was modelled using particle interphase transfer model and Gidaspow drag model on a proprietary computational fluid dynamics (CFD) code ANSYS CFX, using turbulent k-ε model. The interaction between gas and solid particles was modelled using particle interphase momentum transfer model with Miller-Gidaspow drag model on the commercial code. The simulation was run on steady state and then on transient conditions.

Computational fluid dynamics (CFD) usefulness for FCC riser reactor is reaching an advanced stage in solving real life problems. It is a good practise to run a time-average comparison for solid velocities and concentrations to measure fluxes and densities along the riser axes. However, the hydrodynamic behaviour of gas-solid flow in riser was successfully compared with published experimental data of a CFB/FCC riser of length 15.1m and 0.1 m diameter. The overall flow patterns within the riser bed were predicted well by the model. For volume fraction around 2-3%, which is the average particle concentration in the riser system, the computed solid-holdup agrees with the experimental measurements. The predicted results and analysis will be useful for further modelling of industrial FCC riser reactors.

Keywords: CFD, FCC, kinetic theory, hydrodynamic, two-phase, modelling, industrial application

1. Introduction

Circulating fluidised bed (CFB) reactors are widely used in the chemical and process industry to carry out a variety of multiphase gas-solid reactions. The extended use of fluidised bed equipment has opened wide possibilities for ensuring reliable design and improving various industrial technologies, which include: coal combustion and gas-solid catalytic reactions (gasification) such as fluid catalytic cracking (FCC) of long chain hydrocarbons, Fischer-Tropsch synthesis, maleic anhydride production, polymerization of olefins, coal gasification, incineration of solid waste, acrylonitrile production etc.

Fluid catalytic cracking (FCC) is a multibillion dollar worldwide industrial operation to converts heavy hydrocarbons (petroleum) to lower and more valuable molecular-weight products, such as gasoline. In the riser of a CFB, gas is passed through a bed of granular solid material at high velocities, which entrains the solids from the fluidised bed and transports them out of the riser. Based on the bed structure and solid distributions along the riser height three distinct fluidisation regimes have been identified [1], namely: turbulent fluidisation, fast fluidisation and pneumatic conveying. Most industrial CFB reactors operate in the turbulent and fast fluidisation regime [2,3]. These fluidisation regimes are not fully understood [2, 4] and phenomenological flow models are not yet well established [1], consequently the reactor performance predictions are uncertain. Therefore there is a need for an enhanced understanding of the hydrodynamics of CFB in order to facilitate the reactor design and the selection of appropriate operating conditions to achieve the desired fluidisation regime. With recent advances in computational fluid dynamics (CFD) for multiphase flow, this approach can be used for the investigation of complex flow in CFB risers in a cost-effective way. However, such models require detail validation against data from large-scale experimental rigs before they can be used with confidence for the above purposes.

In the present study, gas-solid flow behaviour in the riser system of a CFB has been simulated using a proprietary CFD code [5]. The simulation is based on the experiments carried out by Huang et al. [6] in a pilot-scale FCC riser with 15.1m in height and 0.10m in diameter, which provided a sufficiently long distance for flow development. The solid used in the riser was FCC particles of 67μm mean diameter having density of 1500kg/m^3. The solid mass flow rate was varied between 50-200kg/m^{2}s. Air at ambient conditions was used in the experiment with superficial velocity varying between 3.5-8.0m/s. Figure 1 represents the schematic of the of the riser section of the CFB [6] simulated in this study. The CFD modelling methodology, embedded in the ANSYS CFX code [5], used in the calculation is based on the Eulerian-Eulerian multiphase flow model with constitutive equations for the gas-solid drag, solid stress, solid pressure, solid shear and bulk viscosity, and granular temperature. The solid pressure, shear and bulk viscosity were obtained in terms of granular temperature using the constitutive equation [5] derived from the kinetic theory of granular flow. Modelling of the hydrodynamics of gas-solid multiphase systems with Eulerain-Eulerian models using different CFD codes has shown the suitability of this simulation approach for modelling fluidised-bed reactors. The gas phase turbulence was handled using the k-ε model. Full details of the computational procedure are provided in the paper.

The computed results for solid circulation rates of 100kg/m^{2}s (with superficial gas velocities of 3.5 and 5.5 m/s) are compared with the experimental data [6] of radial distributions of particle concentration and mean velocity at 8 different locations along the height of the riser. The measured trends are generally well replaced in CFD predictions. In this work, detail comparisons between the predicted and measured particle velocities along the axis the radius at various height of the riser length measured from the inlet were presented.

2. Hydrodynamic Model Equations

The hydrodynamic model for the multiphase flow is based on the generalisation of Navier-Stokes equations, which uses Eulerian-Eulerian approach [7]. The Eulerian-Eulerian approach is particularly suitable to simulate industrial bubbling/turbulent riser bed reactors (which may contain billions of solid particles). These equations use granular kinetic theory for the particulate phase. The basic fundamental equations given below are the mass, momentum and energy conservation laws for the fluid and the particulate phases. The CFD modelling methodology, embedded in the ANSYS CFX code [5], used in the calculation is based on the Eulerian-Eulerian multiphase flow model with constitutive equations for the gas-solid drag, solid stress, solid pressure, solid shear and bulk viscosity, and granular temperature. The solid pressure, shear and bulk viscosity were obtained in terms of granular temperature using the constitutive equation [5] derived from the kinetic theory of granular flow. This model is used in this study with the modified Wen and Yu drag law, as in [8]
The gas phase turbulence was handled using the k-ε model.

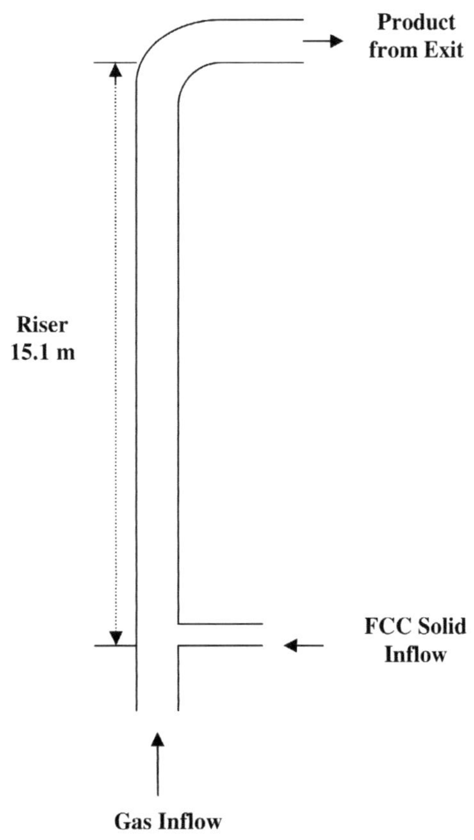

Figure 1 Schematic diagram of a CFB/FCC riser system. The riser is of length 15.1 m and 0.1 m in diameter.

In the 21st century, engineers want to do better still. Why? Because increasing the production of desired product by 1% would increase the profit per reactor by $1 to $2 million/year [1]. Again, due to increase in energy cost and depleting petroleum resources, there is a need for FCC riser optimisation. Equally, the quest is to optimize FCC riser reactor using CFD in order to improve efficiency, cost, environmental impact and health and safety of managing manpower of the process plant. The best use of this commercial software is the flexibility of varying parameters to optimise the simulation solution.

2.1 The governing conservation equations

The continuity equation for gas phase total mass balance is given by

$$\frac{\partial}{\partial t}(\varepsilon_g \rho_g) + \nabla.\left(\varepsilon_g \rho_g \vec{v}_g\right) = 0 \tag{1}$$

Solid phase continuity

$$\frac{\partial}{\partial t}(\varepsilon_s \rho_s) + \nabla.\left(\varepsilon_s \rho_s \vec{v}_s\right) = 0 \tag{2}$$

The volume fraction balance equations is (φ=g, s)

$$\varepsilon_g + \varepsilon_s = 1 \quad \text{or} \quad \sum_g r_g + \sum_s r_s = 1 \tag{3}$$

Where ρ is the density of each phases, ε is the volume fraction, and $\vec{v}$ is the velocity vector

Momentum equations

The mass conservation equations for phase (φ=g, s) is

$$\frac{\partial}{\partial t}(\varepsilon_{\varphi}\rho_{\varphi})+\nabla\bullet\left(\rho_{\varphi}\varepsilon_{\varphi}\vec{v}_{\varphi}\right)=0 \qquad (4)$$

Momentum conservation gas-phase

$$\frac{\partial(\varepsilon_{g}\rho_{g}U_{g}^{i})}{\partial t}+\frac{\partial(\varepsilon_{g}\rho_{g}U_{g}^{j}U_{g}^{i})}{\partial x^{j}}= -r_{g}\frac{\partial P_{g}}{\partial x^{i}}+\frac{\partial(\varepsilon_{g}\tau_{g}^{ij})}{\partial x^{j}}+\varepsilon_{g}\rho_{g}g^{i}+M_{g}^{i} \qquad (5)$$

Momentum conservation solid phase

$$\frac{\partial}{\partial t}(\varepsilon_{s}\rho_{s})+\nabla.\left(\varepsilon_{s}\rho_{s}\vec{v}_{s}\right)+\nabla.\left(\varepsilon_{s}\rho_{s}\vec{v}_{s}\vec{v}_{s}\right)=\nabla\vec{T}_{s}-\beta_{B}.\left(\vec{v}_{s}-\vec{v}_{g}\right)+\varepsilon_{s}(\rho_{s}-\rho_{g})\vec{g} \qquad (6)$$

Where P is the pressure, $\vec{T}_{g}$, $\vec{T}_{s}$ are the stress tensors, β is the interface momentum transfer coefficient, and g is the acceleration due to gravity. Hence, $\vec{v}_{gs}$ is the interphase velocity and $\varepsilon_{s}(\rho_{s}-\rho_{g})\vec{g}$ is an interaction force between phases. $\bar{\bar{\tau}}$ is the φ_{th} phase Reynolds stress tensor

$$\bar{\bar{\tau}}=\varepsilon_{\varphi}\mu_{\varphi}\left(\nabla\vec{v}_{\varphi}+\nabla\vec{v}_{\varphi}^{T}\right)+\varepsilon_{\varphi}\left(\lambda_{\varphi}-\frac{2}{3}\mu_{\varphi}\right)\nabla\bullet\vec{v}_{\varphi}I \qquad (7)$$

Hence, μ_{φ} and λ_{φ} are the shear and bulk viscosity of phase φ, respectively, and I is the unit tensor. The solid pressure, P_{s}, solid shear viscosity λ_{s}, are calculated from the kinetic-frictional stress model discussed in the following section.

Constitutive equations

The ideal gas law is applied to determine the gas pressure where T is the temperature, but this is a cold flow simulation where ambient conditions are applied. Then is the gas phase, the pressure of the gas is given as:

$$P_{g}=\rho_{g}\bar{R}T \qquad (8)$$

Assuming a constant density case, the solenoidal total volume flux is given as in [9]

$$\frac{\partial}{\partial x^{j}}(\sum_{g}r_{g}U_{g}^{j})+\frac{\partial}{\partial x^{i}}\left(\sum_{s}r_{s}U_{s}^{i}\right)=0 \qquad (9)$$

Interphase momentum transfer

The drag on single particle expressed in terms of dimensionless drag coefficient is given as:

$$D = \frac{1}{2} C_D \rho_f A_p \left| U_f - U_p \right| \left(U_f - U_p \right) \tag{10}$$

Where C_D is the non-dimensional drag coefficient of the gas-solid flow. Assuming that the solid particles are of fixed diameter, and then the volume of single particle is given as:

$$V_P = \frac{\pi d_p^3}{6} \tag{11}$$

$$S_P = \frac{\pi d_p^2}{6} \tag{12}$$

Then the number of particle per unit volume is given as:

$$n_p = \frac{\text{volume occupied by particles in unit cell}}{\text{volume single particle}} \tag{13}$$

$$\Rightarrow n_p = \frac{r_p}{V_p} = \frac{6 r_p}{\pi d_p^3} \tag{14}$$

And

$$\mathrm{Re}_{\alpha\beta} = \frac{\rho_{\alpha\beta} \left| \underline{U}_\beta - \underline{U}_\alpha \right| d_{\alpha\beta}}{\mu_{\alpha\beta}} \tag{15}$$

2.2 *Kinetic theory of granular flow*

The granular temperature models are based on the interpenetrating continuum assumption. Both gas and solid phases are modelled as a continuum. In this approach, individual particle trajectories are not simulated but an attempt is made to represent physics of those trajectories and particle-particle interactions using averaged form of governing equations. Due to the use of such averaged equations, models based on this approach can be extended to simulate gas-solid flows comprising large number of solid particles. In this industrial code, the algebraic equilibrium approach was computed by the solver using Gidaspow correlation for the radial distribution on the FCC solid particles. The solid shear viscosity, the solid pressure model and the solid bulk viscosity were computed as a function of granular temperature. For granular flows in an incompressible regime (i.e., where the solids volume fraction is less than its maximum allowed value), a solids pressure is calculated independently and used for the pressure gradient term. According to the granular kinetic theory, the kinetic energy of granular mean flow first degrades in the kinetic energy of random particle fluctuations, and then dissipates as heat of inelastic collisions [9].

Following Lun et al's theory, the kinetic energy of fluctuations is accounted for in the granular kinetic theory by defining a granular temperature Θ_s:

$$\Theta_s = \left\langle c^2 \right\rangle / 3 \tag{16}$$

Where c is the particle fluctuating velocity. Thus, the granular temperature for the solid phase is proportional to the kinetic energy of the random motion of the particles.

2.3 *The Algebraic Equilibrium Model*

ANSYS CFX is restricted to models where the granular temperature Θs is determined algebraically. Previous work has shown that direct adoptions of the default Eulerian-Eulerian models available in FLUENT or CFX, the simulation may not give right predictions on hydrodynamics [10]. Effort is on-course in CFX where granular temperature is determined using transport equation model, using the UDF (user-defined function). The Algebraic Equilibrium model has the flaw that unphysically large granular temperatures can be generated in regions of low solid particle volume fraction. To circumvent this, it is recommended that we specify an upper bound for the granular temperature. The Zero Equation Model implements the simpler algebraic model of Ding and Gidaspow [7] was applied in this simulation and is given as:

$$\Theta_s = \frac{1}{15(1-\varepsilon)} d_s^2 \cdot S^2 \tag{17}$$

Eqn (17) is based on the algebraic equilibrium with the rate of production equal rate of dissipation energy, and is represented as follows:

$$\Pr oduction = Dissipation \Rightarrow \tau_{sij} \frac{\delta U_i}{\delta x_j} = \gamma_s \tag{18}$$

Where τ_{sij} denotes the solids shear tensor.

$$\gamma_s = 3(1-e^2) r_s^2 \rho_s g_0 \Theta_s \left[\frac{4}{d_s} \left(\sqrt{\frac{\Theta_s}{\pi}} - \frac{\delta U_k}{\delta x_k} \right) \right] \tag{19}$$

The restitution coefficient r_{coeff} for particle collisions and g_0 is a radial distribution function and can be seen as a measure for the probability of inter-particle contact and expressed by [10]:

$$g_0 = \left[1 - \left(\frac{\varepsilon_s}{\varepsilon_{s,\max}} \right)^{1/3} \right]^{-1} \tag{20}$$

Where $\varepsilon_{s,\max}$ is the maximum particle packing

Recently, [11] showed that this for the radial distribution function agrees with experimental data for FCC particles. Furthermore, in effort to resolve the fluctuating energy equation, we need to specify the conductivity of the particle fluctuating energy, *k*, and the collisional rate of energy dissipation per unit volume, γ

$$\gamma = 3(1-e^2) \epsilon_s^2 \rho_s g_0 \Theta \left[\frac{4}{d_s} \left(\frac{\Theta}{\pi} \right)^{1/2} - \nabla \cdot v_s \right] \tag{21}$$

The gas-solid drag coefficient for different solid concentrations as presented in [11] as follow:

For $\in_g \geq 0.8$:

$$\beta = \frac{3}{4} C_d \frac{\in_s \rho_g |v_g - v_s|}{d_s} \in_g^{-2.65}, \tag{22}$$

where the drag coefficient C_d is given by

$$C_d = \frac{24}{\text{Re}_s}\left(1 + 0.15\,\text{Re}_s^{0.687}\right); \quad \text{Re}_s < 1{,}000 \tag{23}$$

$$C_d = 0.44; \quad \text{Re}_s \geq 1{,}000 \tag{24}$$

$$\text{Re}_s = \frac{\in_g \rho_g d_s |v_g - v_s|}{\mu_g}, \tag{25}$$

For $\in_g < 0.8$:

$$\beta = 150 \frac{\in_s^2 \mu_g}{\in_g^2 d_p^2} + 1.75 \frac{\rho_g \in_s |v_g - v_s|}{d_s \in_g}, \tag{26}$$

2.4 *Interphase momentum transfer*

For an inhomogeneous phase, the interphase momentum transfer, M_{gs}, occurs due to interfacial forces acting on each phase, due to interaction with another phase β. The total force on phase *g* due to interaction with other phases is denoted M_g, and is given by:

$$M_g = \sum_{s \neq g} M_{gs} \tag{27}$$

Note that interfacial forces between two phases are equal and opposite, so the net interfacial forces sum to zero.

2.4.1 *Initials and boundary conditions*

The definition of appropriate initial and boundary conditions is critical and important for carrying out of a realistic simulation. The system conditions studied are those in [6] and are represented in Table 1 below.

1) At the initial condition, the riser column was assumed empty, and the velocity of both gas and FCC particles were set to be 3.5m/s with 100kg/m^{2}s solid loading (first case) and 5.5m/s with 100kg/m^{2}s solid loading (second case).
2) At the outlet, average static pressure of zero was set, based on the atmospheric inlet reference pressure. Particularly important is also the specification of appropriate boundary conditions at the wall (default domain).
3) Along the wall, a no-slip boundary condition was applied and free-slip for particles was used. A smooth wall influence was based on fluid dependence flow.

Table 1 Inlet, outlet and boundary conditions used in the simulation

Parameter	Case I	Case II	Case III*
Inlet conditions:			
Solid mass flux (kg/m^2s)	100	100	200
Solid velocity (m/s)	3.5	5.5	5.5
Gas superficial velocity (m/s)	3.5	5.5	5.5
Air volume fraction (dimensionless)	0.9813	0.9906	0.9763
Solid volume fraction (dimensionless)	0.0187	0.0094	0.0227
Outlet conditions:			
Pressure boundary conditions	0.0	0.0	0.0

*Case III was not reported in this paper

4) An outlet boundary opening is set for the outflow, to prevent overflow using a convenient transient simulation approach.
5) Global initialisation was done based on Cartesian velocity component having values of u=0, v=0 and w=3.5 and the 5.5m/s at reference volume fraction of air.

2.5 Computational domain and scope

The simulations were carried out for the riser section of the circulating fluidised bed shown in Fig. 1. A 3-D ANSYS CFX-Design modeller was used in developing the grids according to some specification from [9]. The system geometry and system properties are defined in Table 1. An unstructured grid was used throughout of a limiting case of a multi-block grid where each individual cell is treated as a block. The advantage of this arrangement is that there is no implicit structure of co-ordinate lines imposed by the grid. Fig. 1. is a three-dimensional Cartesian coordinates system used. The system geometry and system properties are defined in Table 2. In this study, the restitution coefficient of 0.99 was used. Previous studies have given a range of 0.80-0.99. The restitution coefficient is an empirical input in the kinetic theory based CFD model.

Table 2 Experimental conditions in the fluidised bed riser used by [6]

Parameters	Description	Experiments	Simulation
ρ_g (kg/m^3)	Gas density	Not specified	1.185
ρ_s (kg/m^3)	Solid density	1500	1500
μ_g (kg/ms)	Gas dynamic viscosity	Not specified	1.831E-05
μ_s (kg/ms)	Solid viscosity	Not specified	1.00E-05
d_p (μm)	Particle diameter	67	67
r_{coeff} (dimensionless)	Particle restitution coefficient	Not specified	0.99
d_i (m)	Riser diameter	0.10	0.10
h_i (m)	Riser height	15.10	15.10

2.6 *Solution procedure*

The set of governing equations in section 2.1 – 2.4 was solved by finite control volume technique. Miller-Gidaspows' drag model option was chosen in the CFD input code (CFX11.0). The properties of gas and FCC particles such as specific heat, viscosity, are temperature and composition dependent. The coefficient of restitution is an empirical input in the kinetic theory based CFD model. It is obtained by matching the experiment with the theory. A simulation of high coefficient of restitution

Grid dependency study is the best application method that makes numerical simulation complete. In this work, the grid dependence is tested by varying the mesh numbers and the time step was kept to meet the courant number requirement. The typical values of under relaxation factors between 0.06 – 0.27 were set as default in the solver.

Advection scheme of high resolution was set based on steady/unsteady state scheme using second backward Euler simulation. A residual type convergence criterion, RMS having a target of 1.0E-04 was used.

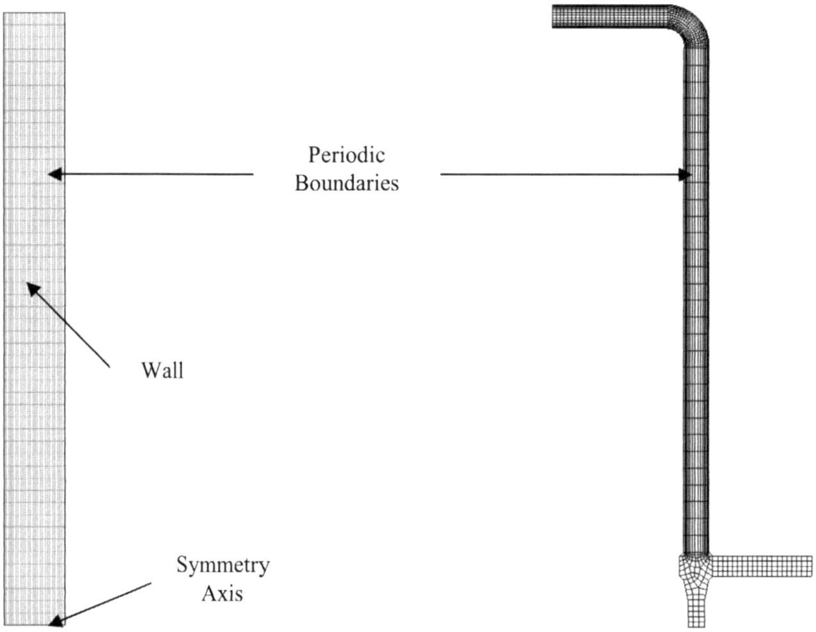

Figure 2. Straight pipe unstructured Mesh

Figure 3. Riser pipe bend Mesh

3. Simulation

The simulation model basically consists of a turbulent flow, where two phases enter into contact: a gas phase formed by ideal gas at 25^0C and a particulate FCC solid phase composed of catalyst properties. The particles are considered smooth, spherical and inelastic, with an average diameter of 67μm and a density of 1500 kg m^{-3}. Table 1 and 2 shows the details of the feed properties input into the riser system.

Boundary conditions

At the entrance, all velocities and concentrations of both phases are specified. The set-up was based on atmospheric condition of temperature and pressure, fluid treated as incompressible. At the initial conditions, the superficial gas velocity is specified at 3.5 m/s (Case 1) and 5.5 m/s (case 2). At the walls, the gas velocity is zero; the FCC particle phase velocity has a no-slip condition.

Mesh and computational code

An unstructured mesh is composed of 156,000 control volumes, 476,690 faces and nodes 164,892. The details of its refinement, both at the entrance and at the exits, are presented in Figs 2 and 3. The time step is on the order of 10^{-4}s. Adaptation of the mathematical model for the numerical model generation was achieved with the ANSYS CFX 11.0 commercial simulator, which is based on the finite volume method, incorporating the high resolution scheme.

In the dependency evaluation of the flow with the numerical mesh, it is known that an adequate number of control volumes are of extreme importance to avoid numerical errors, which is not possible with less refined meshes. Thus, five meshes, with 133,495, 156,000, 596,964, and 985,265 and 1,334,925 control volumes respectively, were tested. The mathematical-default model used in accordance with the practice normally adopted, i.e., the flow inside the riser equipment is purely gaseous (ideal gas), subjected to the conservation equation of continuity and momentum for each phase. The simulations with different meshes resulted in the following observations:

- *Axial dependency* (r = 0 m and z = 0 – 14.08 m): There was no much significant influence on the mesh refinement in the axial direction. However, a noticeable variation was verified in the less refined mesh in the feed contact region, which is considered a critical point in the experimental case studied.
- *Redial dependency* (z = 0.95 m): In this transverse section, practically in the feed contact region, the radial behaviour of the flow is analogous to that described for the axial dependency.
- *Redial dependency* (z = 8.16 m): In this section, an insignificant influence of the refined mesh was also observed; all meshes provided the same qualitative profile and quantitative profile.
- *Redial dependency* (z = 14.08 m): Dependency on the refined mesh was more noticeable toward the exit section, even with clear difference observed. However, this does not modify the qualitative results.

Then the mesh composed of 156,000 control volumes was chosen; this was shown to be in accordance with the established standards (that is, there is no significant difference between the calculated flow of this mesh and those of the refined ones), when compared with the predictions from the other meshes.

4. Results and Discussion

Initially the numerical meshes were tested done and the one that had the best behaviour in the standard flow was used to run the simulations for the real process. The particle velocity radial profiles for the CFB/FCC risers were analysed, using the same conditions from the experimental studies reported. Thus, the report of [11] has show a good agreement between model predictions and experimental data of [12]. Instead of repeating those simulations, in this work, we have simulated gas-solid flow corresponding to experimental conditions of [6]. In order to determine the validity of these models, the simulation results are compared with experimental data using same input variables in the tests as the simulation program input. Quite a number of simulations have been performed in order to investigate the effect of different operating conditions, model assumptions, and to get an accurate description of the observed gas-solid phase flow pattern in the riser. The main interest here is on the solid phase volume fraction distribution, velocity distribution of both phases, granular temperature, FCC viscosity and solid hold-up.

Table 3: ANSYS CFX-Pre input simulation setup

Model	Parameters	Values and Comment
	Simulation Type:	Steady state/Transient Simulation
Fluid:	Fluid list:	Air at 25^0C and FCC Particles
	Reference pressure:	1-atm
	Buoyancy option:	Buoyant
	Domain motion:	Stationary
	Fluid models:	*Eulerian-Eulerian* Non-homogeneous model
	Turbulence model:	k-ε model
	Fluid details:	Ideal Air and FCC Particles Air at 25^0C as continues fluid FCC Particles as dispersed solid
	FCC Particle mean diameter:	67μm
	Restitution coefficient:	0.99
	Kinetic Theory:	Granular Temperature Model: *Algebraic equilibrium*
	Model option used	Miller-Gidaspow
	Radial Distribution Function:	Kinetic Theory
	Solid Pressure Model:	Kinetic Theory
	Solid Bulk Viscosity:	Kinetic Theory
	Solid Shear Viscosity:	Kinetic Theory
	Interphase transfer:	Particle Model
	Drag force used	Gidaspow
Fluid Default:	Wall influence on Flow:	Fluid dependent
	Wall Roughness:	Smooth wall
	Wall contact model:	Volume fraction used
	Air at 25^0C:	No slip
	FCC Particles:	No slip/Free slip
Fluid Inlet	Fluid Region:	Subsonic: Normal Speed: 3.5 and 5.5 m/s Solid Concentration: 100 kg/m^{2}s
	Air volume fraction:	0.9813 (dimensionless) at 3.5m/s
	FCC Particle volume fraction:	0.0187 (dimensionless) at 3.5m/s
	Turbulence Intensity:	Medium: 5%
Fluid Outlet (Opening)	Fluid Region:	Subsonic:
	Mass and momentum:	Average Static Pressure: 0.0Pa
	Turbulence:	Zero gradient
Solver control:	Discretization (Advective) scheme:	High Resolution
	Transient scheme:	Second order Backward Euler
	Convergence Criteria:	RMS
	Residual target:	1.0E-04
	Equation class setting:	Continuity
	Time step (Residual Target):	1.0E-4s
	Total Time:	40-secs

4.1 Gas phase distribution (single phase)

The initial setup case was based on gas (single) flow in the riser domain. The results were to prove the capability of the solver code in term of the turbulence modelling of the flow. The results were found to be in agreement with the 1/7th power law profile as observed by [12] and [13]. Fig 4 and 5 show the time-mean single phase turbulent velocity profile achieved in the initial set case. At the inlet axis of the riser, there was a uniform velocity flow which subsequently developed to turbulent flow regime [14]. Fig 5 shows the further development of the flow into a fully developed regime. The riser is of total length 15.1m, but the velocity comparison with the experimental case was based on the probe point of 14.08m. These flows were compared favourably with the $1/7^{th}$ power law profile based in basic fluid dynamic flow [15].

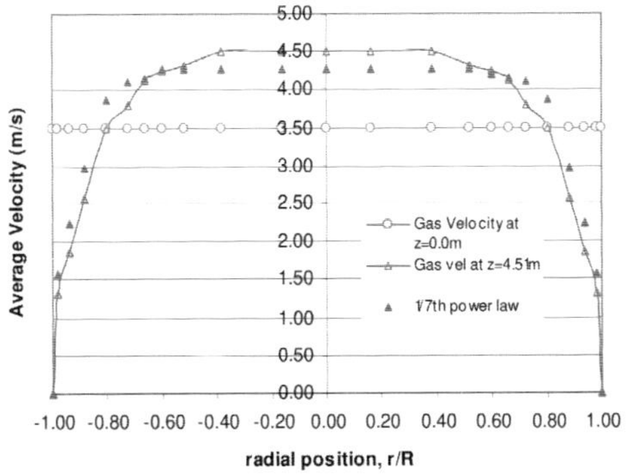

Fig 4. Gas velocity profile at z =0.0m and 4.51m

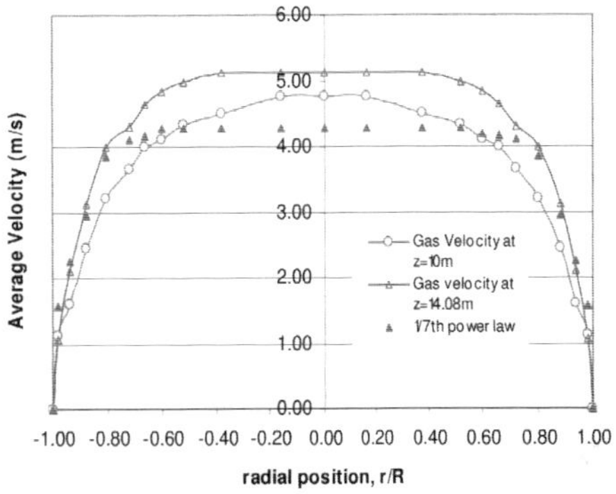

Fig 5. Gas velocity profile at z = 10m and 14.08m

Fig 6 shows axial profiles of cross-sectional average solids holdup in both the 15.1 m and 10 m heights. In this figure, its vivid that the distribution in the riser are uniform except at the inlet region. The solid distribution at the entrance region is highly influenced solid structure. The fully developed region in the 10 m riser is much lower than that of the 15.1 m. Thus, the riser height is of great importance to the solids distribution. From Fig. 6, it's clear that the decrease in the superficial gas velocity results to increase in the cross-sectional average solids holdup, while increasing solids fluxes increases the cross-sectional average solids holdup which has been shown in previous studies [15, 16].

4.2 Axial Distribution of Solids Holdup in the FCC riser

Fig. 6 represents the axial distribution profile of cross-sectional solids holdup in the CFB/FCC riser of height of 15.1 m. A summary of this riser is shown in Fig. 1. I n the simulation setup case, the solids circulation rate was set to be 100 kg/m^2s on a varying superficial gas velocity between 3.5 m/s and 5.5 m/s. It is very clear from Fig. 6, that there is uniformity in axial distribution in the riser system. The non-uniformity is shown at the entrance region of the riser. The factor that influences the solids distribution is basically the distribution structure, which has been seen in previous studies [6]. Therefore, the riser height is important to the solids distribution in CFB/FCC riser system. From Fig. 6, it is shown that decreasing the superficial gas velocity result an increase in cross-sectional average solids holdup and vice versa [6].
Figs. 7 and 8 represent the contour plot at the upper-end of the riser system. Fig. 8 show the vector plot of the flow while Fig. 9 represent the contour plot of the injection-mixing section in the riser system.

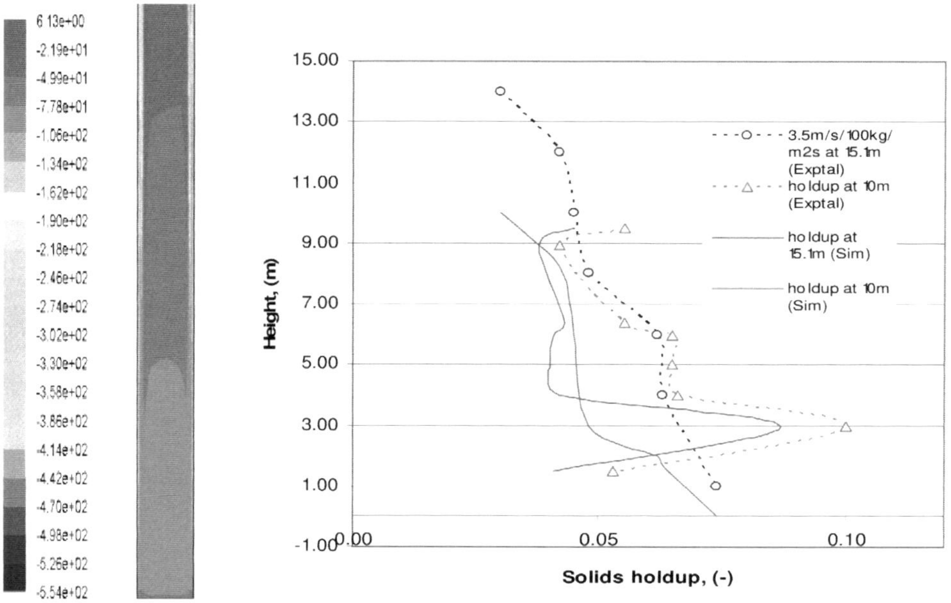

Fig 6. Comparison of axial solids holdup at various heights with contour plot of FCC Particle flow

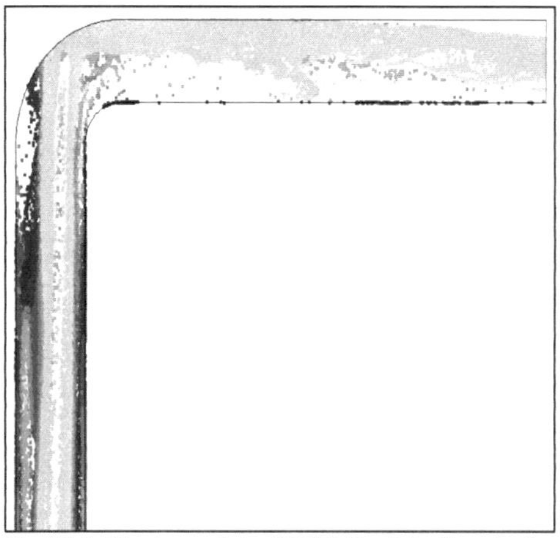

Fig 7(a). Contour plot of FCC Particle flow in the riser

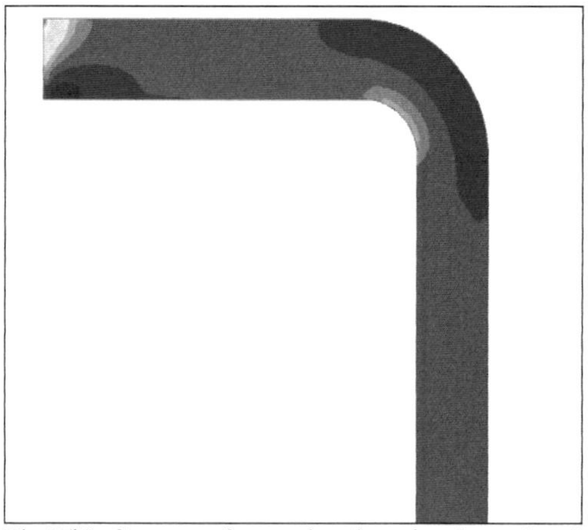

Fig.7(b). Contour of upper bend section (gas-phase)

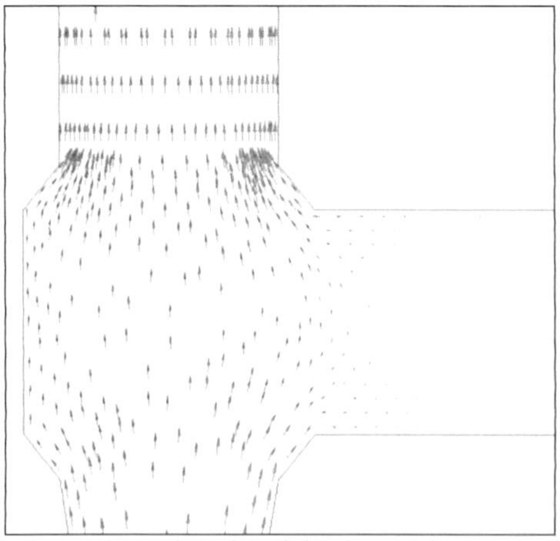

Fig 8. Vector plot of the injection-mixing section

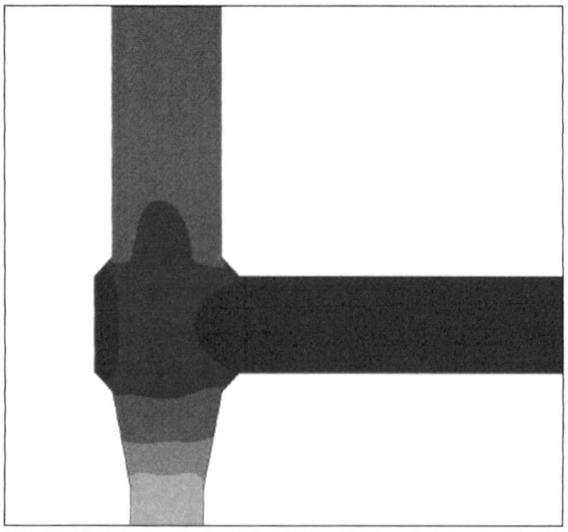

Fig.9. Contour plot of the injection-mixing section

4.3 FCC Solid Velocity distribution of both gas-solid phases

In developed flow, the slip velocity is approximately the terminal velocity. For the Geldart A particles (e.g. FCC particles at 1500kg/m^3), the solid velocity is close to the gas velocity in the direction of flow. A balance between buoyancy and drag obtained from the basic momentum balances for one-dimensional, developed flow is as follows:

The simulations carried out throughout this research showed that in each one of the transversal sections, there are three distinct zones: a dilute central zone (i.e., a zone where the particle concentration is very low, and consequently its velocity is high, that extends from the radial centre ($r = 0$) to approximately $r = 0.2\ m$; an intermediate zone with a moderate particle velocity (approximately between r = $0.2\ m$ and $0.4\ m$); and a wall zone with a low particle velocity that extends from approximately $r = 0.4\ m$ to $1.0\ m$.

Fig 6 shows the time-mean two-phase vertical velocities at riser length z = 0.95m. The velocity profile was over-predicted and the downward flow of the particles along their carrier gas near the wall is predicted, although the no-slip boundary will guarantee a zero value for gas velocity at the immediate wall position. Fig 7(a) show the contour plot of FCC solid particles in the riser system. While Fig 7(b) represents the contour plot of single (ideal gas) phase flows including the upper and bend of the riser system. The nature of the flow field and the granular flow velocity can be seen from the contour plot. Figs. 10-11 shows the profiles of velocity in the dense region (z = 0.95 m and 2.59 m) of high solids concentration. It can be observed that inviscid model has quantitative behaviour in the constant coefficient Newtonian model, but this behaviour is inverted closer to the wall. This is also in good agreement with [1]. The deviation comparison of the predictions and experimental study observed in Fig 8 is about 0.05%.

Fig 11 show the under-predicted FCC solid velocity profile when compared with the experimental data. By observation, it can be seen that the computation was almost at the fully developed flow level between the plane z = 0.95 m and z = 2.59 m. Figs 12 and 13 show a more pronounced under-predicted model when compared with experimental result [1]. The deviation is about 5.2% at this region. Since various issues like: the grid dependency, turbulent model, discretization scheme and power law equation was resolved. The discrepancies might be on CFD error and/or error from the experimental data used. The model predicted has shown a uniform distribution of FCC solid particle along the length of the riser. The numerical results are in good agreement with the experiment, both in form and magnitude.

4.3.1 Case 1: $Ug = 3.5m/s$ at $Gs = 100\ kg/m^2s$

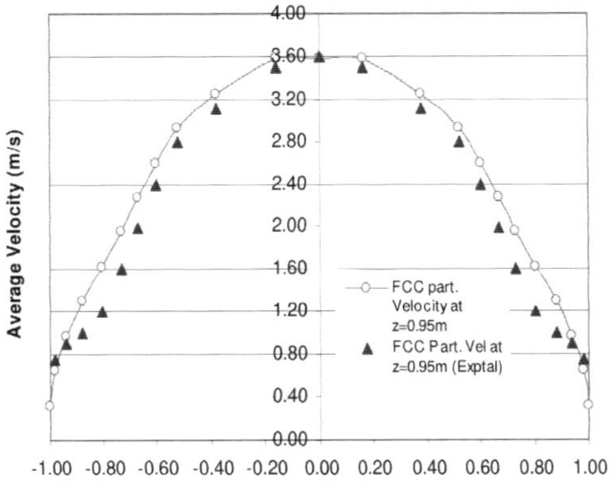

Fig 10. FCC Particle velocity profile at z =0.95m

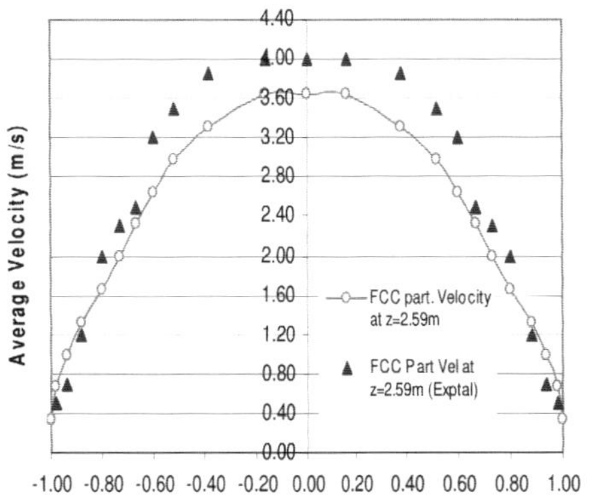

Fig 11. FCC Particle velocity profile at z =2.59m

The solids flow can be considered as fully developed if the radial solids distribution no longer changes with respect to the axial location. With this principle, it is seen that increasing the superficial gas velocity accelerates the flow development in Figs 12, 14, 16 and 18. Again, when the solid circulation rate was increased form 3.5 m/s at 100 kg/m^{2}s to 5.5 m/s at 100 kg/m^{2}s, the flow development are lower and the simulation comparison with the experimental case can be seen in Figs. 13, 15, 17 and 19. In this figures, the predictions of the local solid volume fraction are successful. The reported profile in this study are at different riser heights: 0.95 m, 4.51 m, 10.0 m and 14.08 m. respectively. At all the bed heights in this Figs. 13, 15, 17 and 19, the core–annulus flow were closely predicted. This comparisons also provide the model validation axially [17].

Fig 19 shows a symmetrical particle velocity profile; this is a proof of transient solution to the simulation case. But when the transient simulation was carried out, the results obtained was a constant values at all level of flow in the riser system. At this point we can conclusion that the simulation was a transient case as initially viewed to be steady state. But furthermore there is a need to incorporate a user-define subroutine to address the discrepancies from the default setting. Since the phenomenological model 535 centipoises solid viscous was applied, effort will be is place to look at the viscosity predictions. Again, as the level of dilution is high, the kinetic theory of granular flow (KTGF) is underpredicting the collisional viscosities, hence the flatter velocity profile. This indicates a modelling deficiency of the KTGF in the dilute limit, which will improve when we consider larger solids loadings.

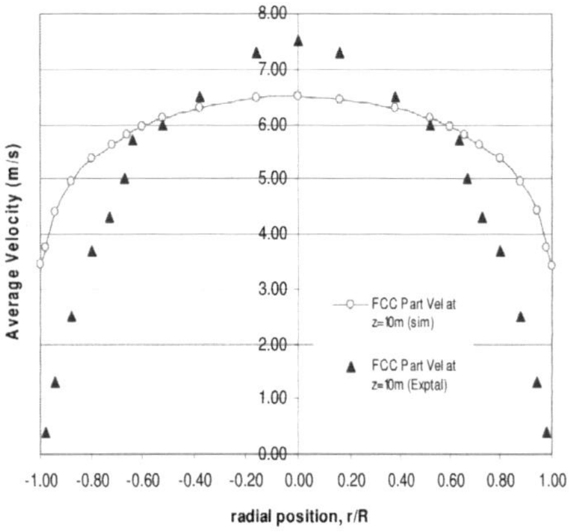

Fig 12. FCC Particle velocity profile at z =10m

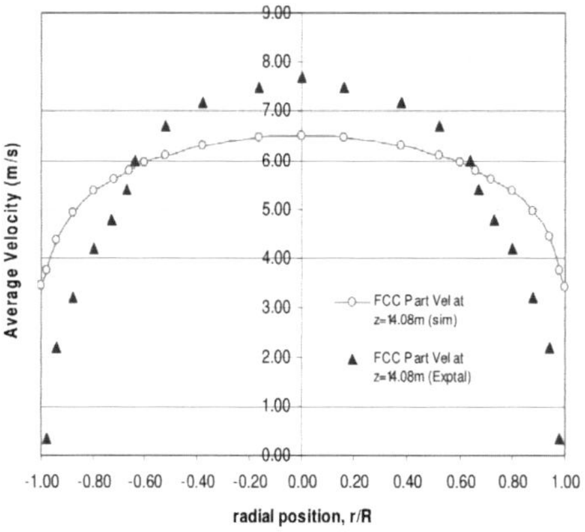

Fig 13. FCC Particle velocity profile at z =14.08m

4.4 *Solid phase volume fraction distribution*

It has been seen in Figs 14, 16, 18 and 20 that the solids concentration in the riser centre remains almost constant throughout the riser under each operating condition. These figures shows that even with the change in operating conditions, the solids concentration seem to remain unchanged within the operating conditions. Therefore, the flow development in the riser mainly depends on the solids distribution in the wall region. Figs 14 and 16 show a close comparison with the experimental data, but the effect toward the riser wall is a little higher than that of experimental data. Fig 19 shows good comparison of a symmetrical profile of solids holdup. Also Fig 21 shows a close and more in symmetric comparison of volume fraction when compared with the experimental data [6]. The wall effect is much noticed to be better predicted at the plane z = 10.0 m and z = 14.08 m. This may be as a result of fully developed flow along the riser length at that region.

Detail knowledge and understanding of the solids distribution is very vital for the design, development and operation of CFB/FCC riser reactor. Gas and solid particle residence time distributions in the riser, the rate of heat transfer and erosion rates of surfaces in fluidised riser beds also depend on solids volume fraction distributions. The solid particles are entrained up in column, against the gravitational acceleration, by the upflowing gas phase [17].

4.3.2 *Case 2: Ug = 5.5m/s at Gs = 100 kg/m²s*

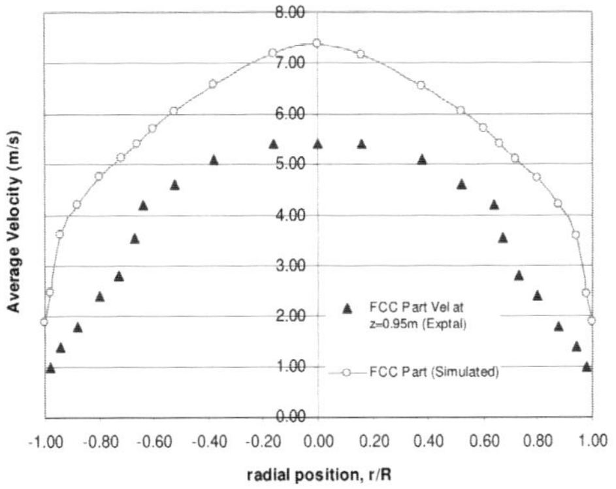

Fig 14. FCC Particle velocity profile at z =0.95m

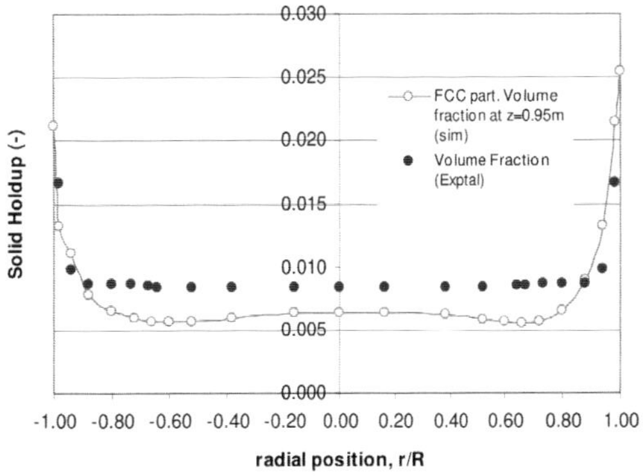

Fig 15. FCC Volume fraction at z = 0.95m

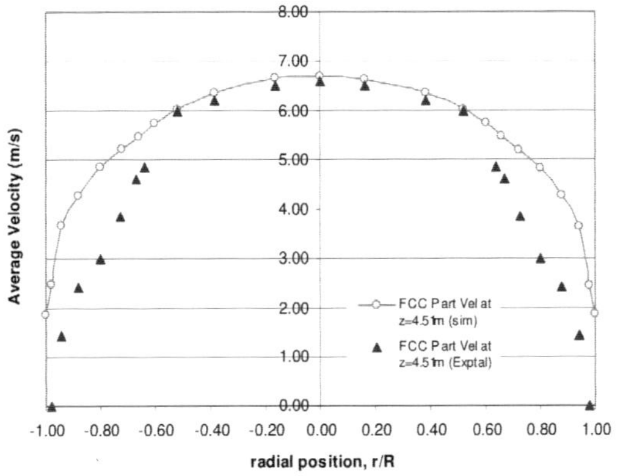

Fig 16. FCC Particle velocity profile at z =4.51m

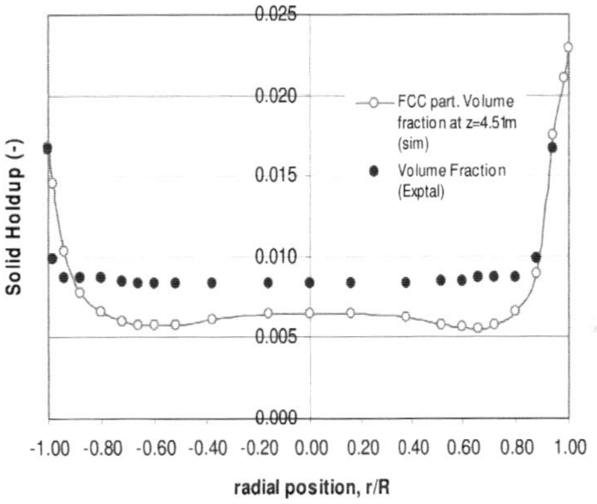

Fig 17. FCC Volume fraction at z = 4.51m

4.5 *Influence of wall restitution coefficient*

In this simulation, it is possible to investigate the influence of the restitution coefficient of the wall on the simulation results. But the default settings used is not showing enough boundary effect toward the wall of the riser. The knowledge of restitution coefficient of the wall is therefore necessary for a correct application of the model. [12] estimated the restitution coefficient of the wall from measurement of granular temperature and solid volume fraction in the dilute regime of a CFB/FCC riser.

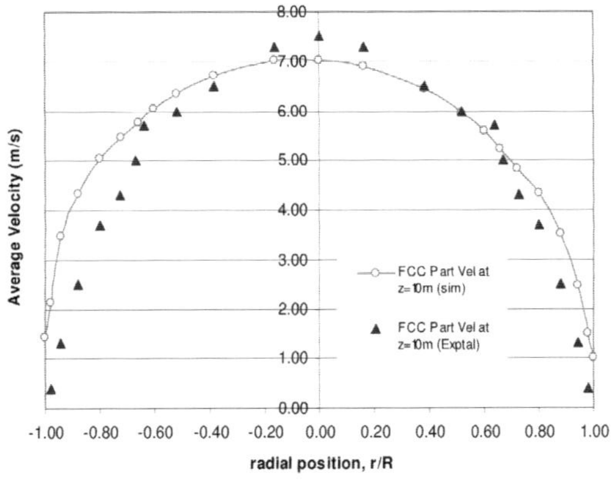

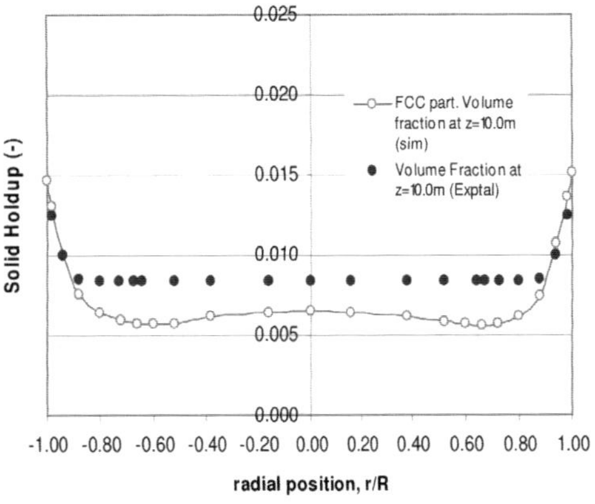

Fig 18. FCC Particle velocity profile at z =10m

Fig 19. FCC Volume fraction at z = 10m

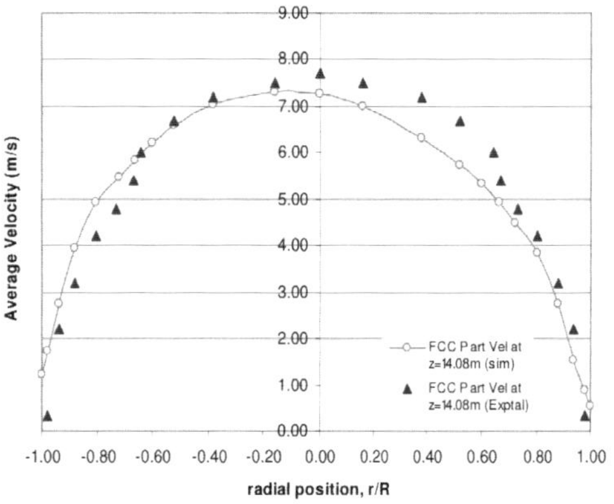

Fig 20. FCC Particle velocity profile at z =14.08m

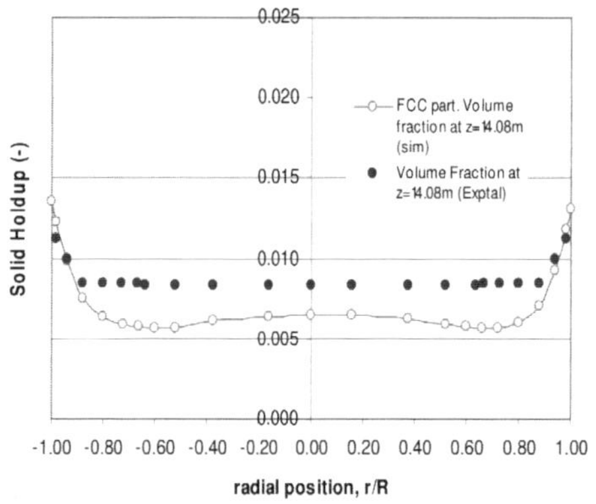

Fig 21. FCC Volume fraction at z = 14.08m

4. Conclusion

A computational fluid dynamics study was done using a commercial code ANSYS CFX 11.0 to describe the hydrodynamic behaviour of gas-solid flow in a FCC/CFB riser. The model was based on Eulerian-Eulerian two-fluid modelling approach, incorporating the kinetic theory constitutive model for a dilute assemblies of the particulate solid, a Gidaspow's drag model for gas-particles interaction. The simulation model takes into account the axial and radial distribution of voidage, velocity and pressure drop for gas and solid phase, and solids volume fraction and particle size distribution for solid phase. The typical flow patterns of the CFB/FCC bed were obtained and compared favourably with the reported experimental results. This model results are compared with and validated against atmospheric cold bed CFB units' experimental data given in the literature for axial and radial distribution of void fraction, solids volume fraction and particle velocity, total pressure drop along the riser height and radial solid flux. The discussions of results were based on the predicted numerical results of gas-solid behaviour in the riser simulation using default setting. The kinetic theory of granular flow model applied in this simulation based on default settings, increases the instability of the Eulerian-Eulerian calculation.

The trends of the flow development are observed from axial and radial solids concentration and particle velocity distributions. For the setup operating boundary conditions, the solids concentration in dilute phase are relatively constant at the riser centre throughout the FCC riser. This is why the argument for the quick solids flow development in the riser centre. However, the flow development in the wall region is quiet slow with the solids holdup near to the riser wall decreasing slowly toward the riser height.

This discrepancies is not only explained by the particle-based approach used in this model, because of the consideration of cluster effect based on drag coefficient, this situation is initiated by the fact that the solids are accelerated to an upward velocity, and there is a very large voidage gradient in the riser bottom.

More so, increasing the overall solids circulation rate, Gs, decreasing the superficial gas velocity, Ug, increases the solids holdup in the wall region throughout the FCC riser length. In conclusion, the simulation results are in close comparison with the experimental data, although there are noticeable under-predictions and deviations which may be corrected using applied user-subroutines in the ANSYS CFX software. The overall flow patterns within the riser bed were predicted well by the model. For volume fraction around 2-3%, which is the average particle concentration in the riser system, the computed solid-holdup agrees with the experimental measurements.

Notation

c	Fluctuating velocity, m/s
C_D	Drag Coefficient
d_p	Particle diameter
g	gravitational constant, 9.81m/s^2
P	Fluid pressure
P_s	Particle pressure
$\bar{R}$	Gas constant, J/mol.K
Re	Reynolds number
r_{coeff}	Coefficient of restitution
$\vec{R}_{gs}$	Drag force between gas and particulate phases, N/m^3
t	time, s
$\vec{v}_g$	The gas velocity, m/s
$\vec{v}_g$	The gas velocity, m/s
$\vec{v}_s$	The solid velocity, m/s
$\vec{v}_{gs}$	The drift velocity, m/s
$\vec{v}_\varphi$	Velocity of phase φ, m/s
x, y	radial coordinates, m
z	axial coordinate, m

Greek letters

e_g	Gas volume fraction
e_g	solid volume fraction
e_φ	volume fraction of phase φ
$\varepsilon_{s,\max}$	maximum volume fraction of particle
Θ_s	Granular temperature, m^2/s^2
λ_s	shear bulk viscosity
μ_g	gas dynamic viscosity, kg/ms
μ_s	Solid shear viscosity, kg/ms
ρ_g	Gas density, kg/m^3
ρ_s	Solid density, kg/m^3
ρ_φ	Density of phase φ, kg/m^3

Subscripts

g	gas phase
k	turbulence kinetic energy
max	maximum
s	solid phase
CFB	Circulating fluidised bed
FCC	Fluid catalytic cracking riser
KTGF	Kinetic theory of granular flow

Acknowledgement

The authors would like to acknowledge the support from the Petroleum Technology Development Fund (PTDF), Abuja, Nigeria, for the sponsorship of this Ph.D. research study at the University of Leeds, United Kingdom and to present this paper at the 100-years centenary of AIChE in Philadelphia, USA.

References

1. Levenspiel, O. (1999) *'Chemical Reaction Engineering'* 3rd Edition John Wiley & Sons
2. Jiradilok, V., Gidaspow, D., Damronglerd, S., Koves, W. J., and Mostofi, R., (2006) *'Kinetic theory based CFD simulation of turbulent fluidization of FCC particles in riser'* Chemical Engineering Science 61, 5544-5559
3. Grace, J. R., (2000) *'Reflections on turbulent fluidization and dense suspension upflow, Powder Technology'* 113, 242-248
4. Du, B., Warsito, W., Fan, L. S., Bed, (2003) *'Nonhomogeneity in turbulent gas-solid fluidization'*, A.I.Ch.E Journal 48, 1896-1909
5. ANSYS CFX (2006) *Solver Theory Guide.* ANSYS CFX Release 11.0: 1996 ANSYS Europe, Ltd
6. Huang, W., Yan, A., and Zhu, J., (2007) *'Hydrodynamics and flow development in a 15.1m Circulating Fluidised Bed Riser'*, Chem. Eng. Techno. 30, No. 4, 460-466
7. Ding, J and Gidaspow, D., (1990) *'A bubbling fluidisation model using theory of granular flow'* AIChEJ. 36 pp. 523 – 538
8. Yang, N., Wang, W., Ge, W., Wang, L., Li, J., (2004) ,*Simulation of heterogeneous structure in a circulating fluidised-bed riser by combining the two-fluid model with EMMS approach.* Industrial and Engineering Chemistry Research 43, 5548-5561
9. Versteeg H. K., and Malalasekera W., (2007) *'An Introduction to Computational Fluid Dynamics: The Finite Volume Method'* 2nd Ed., Pearson Prentice Hall, Harlow, England.
10. Yi Cheng et al (2008) *Downer reactor: From fundamental study to industrial application,* Powder Technology 183: 364-368
11. Gidaspow, D., (1994) *Multiphase Flow and Fluidization: Continuum and Kinetic Theory Descriptions,* Academic Press, New York
12. Gidaspow, D., AND l. Huilin, (1998) *'Equation of State and Radial Distribution Function of FCC Particles in a CFB'* AIChE J., 44, 279
13. Ranade V. V (1999) *Modelling of Gas-Solid Flow in FCC Riser Reactors: Fully Developed Flow,* 2nd International Conference on CFD in Materials and Process Industries, CSIRO, Melbourne, Australia.
14. Yang, Y. I., (1991) .,*Experimental and theoretical studies on hydrodynamics in cocurrent upflow and downflow circulating fluidized beds*, PhD. Diss., Tsinhua University, Beijing, China.
15. Derouin C, et al (1997) *Hydrodynamics of Riser Units and Their Impact on FCC Operation.* Ind. Eng. Chem. Res. 36, 4504-4515
16. Martin, M. P et al (1992) *Catalytic Cracking in Riser Reactor: core annulus and elbow effects.* Chem. Eng. Sci. 47, 2319
17. Gungor A., AND Eskin N., (2007) *Hydrodynamic modelling of a circulating fluidised bed' Powder Technology* 172 p1-13

761d Dynamics and Control of Air Separation Unit-Gas Turbine-Gasifier Integrated Power Cycle of Igcc Plants

Priyadarshi Mahapatra, Rensselaer Polytechnic Institute, Department of Chemical Engineering, 110 Eighth St., Troy, NY 12180-3590 and B. Wayne Bequette, Chemical and Biological Engineering, Rensselaer Polytechnic Institute, 110 8th Street, Troy, NY 12180.

Integrated coal gasification and combined cycle (IGCC) power plants have emerged as an attractive alternative to conventional pulverized coal (PC) plants for generating power from coal. If carbon dioxide capture and storage is required, IGCC is considered one of the most promising power generation technologies. In these plants, the subsections closely interact with each other, resulting in challenging operation and control problems. Examples of process interaction include: (i) the syngas fuel from gasifer serves as feed to the gas-turbine/compressor, (ii) pure oxygen from the air separation unit (ASU) serves as the oxidant for gasification, (iii) some of the compressed air from the gas compressor (which has a common shaft with the gas turbine) serves as feed to the ASU, and (iv) the nitrogen from the ASU is injected into the gas turbine's combustor. The substantial process integration, which is targeted to achieve lower emissions (cleaner coal) and increased efficiency, leads to higher investment costs, reduced operational flexibility and longer startup-shutdown time compared to PC plants.

This paper addresses control of ASU-GasTurbine-Gasifier power cycle by first detailing the development of an AspenDynamics model consisting of individual ASU, Gasifier and GasTurbine subsections. Thereafter these subsections are coupled to investigate feasibility, applicability and plant performance with varying coal quality and with changes in coal-feed quantity required to meet power demands which act as overall process disturbances. The operability characteristic of the integrated process is compared with that of the individual operating units. The process variables operating at different time scales are shown to be responsible for difficulties with commonly used PID-based control schemes. Further, we develop a model predictive control strategy that handles rate-of-change and multiple time-scale constraints imposed by the process design of the air separations, gas turbine/compressor and gasifier units. Finally we close with a discussion of future work on the simulation and control of an entire IGCC power plant.

761e Improving Catalytic Hot Gas Cleaning in Biomass Gasification by Oxygen Addition

Lina Ma, Joeri F.M. Denayer, and **Gino V. Baron**. Department of Chemical Engineering, Vrije Universiteit Brussel, Pleinlaan 2, Brussels, B-1050, Belgium

The success of biomass gasification technology (certainly if combined with Fisher Tropsch and other technologies) depends on the availability of simple, robust and inexpensive gas cleaning methods to eliminate the contaminants such as tars, particles, nitrogen compounds and alkali metals from fuel gas to avoid downstream operating problems.

In our previous work [1], a novel catalytic candle filter was developed to integrate the high temperature removal of particles and tars from biomass gasification gas in a one-step gas-cleaning process in order to reduce the equipment inventory and meet the specifications demanded by end-use devices and environmental constraints. It consists of a ceramic á-alumina porous filter substrate in which a suitable nickel-based catalyst is deposited onto the pore walls to destroy the tars. In the laboratory scale, a small filter disk with a diameter of 29mm to 30mm and a thickness of 10mm is used to represent the full-scale candle filter. After a large amount of development and optimization work, an optimal catalyst composition had already been found, which approximately consists of 2.5 wt% secondary supports (Al2O3, ZrO2 or a mixture of ZrO2-Al2O3), 1.0 wt% Ni and 0.5 wt% MgO in the porous, preformed alumina filter material with low specific surface. In typical conditions with a face velocity of 2.5 cm/s, the presence of 100 ppm H2S and at 900 °C, the conversion of naphthalene is 100%. At 850 °C and the same conditions mentioned above, the naphthalene conversion is about 99.0%. However, when reducing the temperature to 800 °C, only 77% reduction in tars is achieved. This means that this catalytic gas cleaning process needs further improvements to reduce the required reaction temperatures, significantly reducing the equipment cost.

Recently, some groups [2] introduced air or oxygen into the biomass gasification gas to avoid carbon deposits or increase temperature by partial combustion, which resulted in better tar removal and stable operation. High conversions of tars (e.g. toluene) were obtained in the presence of 1% to 3% oxygen below 600°C with the zirconia-containing catalysts, whereas the tar conversions are negligible without oxygen in the feed [3]. To identify the role of thermal oxidation in removing the tars, experiments were carried out using a blank alumina disk with the 1vol% O2 in the gas mixtures: in typical conditions (900 °C, 100ppm H2S, 2.5 cm/s) with 1 g/Nm3 naphthalene as tar model compound, the conversions of naphthalene are 54.0% with 1% O2 and 9.9% without O2. Additionally, at 800 °C, the conversions of naphthalene are 47.3% with 1% O2 and 2.0% without O2. At high temperature, O2 reacts with naphthalene and result a high tar conversion in comparison to the decomposition in absence of O2, where the naphthalene conversion was negligible.

However, with the optimal catalyst compositions, the conversions are much higher and show that the addition of a small amount of oxygen in the biomass gasification gas increases the tar conversion significantly. For example, even as low as 800 °C, the naphthalene conversions are 98.2% with 1% O_2 versus 79.5% without O_2 in the feed. Conversion with only the secondary support is only 70% at 800°C, higher than for the blank material, as ZrO_2 is known to be a catalyst for this conversion in the presence of O_2 [3]. These improvements are clearly linked to the catalytic conversion and cannot be explained on the basis of the thermal conversion only. The addition of O_2 also profits the reforming of CH_4 to make synthesis gas and enhances the ratio of H_2/CO. Long term experiments were performed at 800°C over 250 hours with 1g/Nm^3 naphthalene and 1% O_2. The conversion of naphthalene was ranging from 98.2% to 99.4%, while the pressure drop remained stable over 250 h. Adding a small amount of oxygen to the produced fuel gas allows almost complete removal of tars at low temperature without significantly modifying the gas composition and resulting in significantly simpler and less expensive equipment.

We also believe to have elucidated the mechanism of improvement with the help of a large number of experiments with different catalyst and gas compositions and thermodynamic calculations.

REFERENCES

1. Ma, L., H. Verelst and G.V. Baron, Catalysis Today, 105, 3-4 (2005).

2. Corella, J., J. M. Toledo and R. Padilla, Ind. Eng. Chem. Res., 43, 10 (2004).

3. Juutilainen, S. J., Pekka A. Simell and A. Outi I. Krause, Applied Catalysis B: Environmental, 62, 1-2, (2006).

AUTHOR INDEX

AUTHOR INDEX